Every year, extreme climatic problems occur around the globe, with droughts in some places and floods in others. Recently, we have come to recognize that some of these widely dispersed climatic extremes might have a common origin in the periodic warming of sea surface water in the central and eastern Pacific Ocean. A century ago, Peruvians connected this appearance of warm water every December to changes in environmental conditions and christened it "El Niño", the Spanish term for the Christ Child. In some years, El Niño lingers, and seems to be connected with droughts in Australia, Brazil or India, reduced incidence of tropical hurricanes on the east coast of the USA, and floods in Peru. *Currents of change* explains in simple terms what El Niño is, how its effects might be forecast and its far reaching impacts on all of us.

Currents of change

Currents of change: El Niño's impact on climate and society

MICHAEL H. GLANTZ

*Program Director, Environmental and Societal Impacts Group,
National Center for Atmospheric Research, Boulder, CO*

Published by the Press Syndicate of the University of Cambridge
The Pitt Building, Trumpington Street, Cambridge CB2 1RP
40 West 20th Street, New York, NY 10011-4211, USA
10 Stamford Road, Oakleigh, Melbourne 3166, Australia

First published 1996

Printed in Great Britain at the University Press, Cambridge

Typeset in Times 10/12pt

A catalogue record for this book is available from the British Library

Library of Congress cataloguing in publication data
Currents of change : El Niño's impact on climate and society / Michael H. Glantz
   p. cm.
Includes index.
ISBN 0 521 49580 6 (hbk)   ISBN 0 521 57659 8 (pbk)
1. El Niño Current. 2. Climatic changes. I. Glantz, Michael H.
GC296.8.E4C87 1996
551.47'6 – dc20 96–13992 CIP

ISBN 0 521 49580 6 hardback
ISBN 0 521 57659 8 paperback

Dana Thompson and I were postdoctoral fellows together at the National Center for Atmospheric Research in 1974. He came from the world of physical science, and I came from the world of social science. Dana introduced me to the El Niño phenomenon, and for several years we joined efforts to address the physical and societal aspects of this important recurring natural process. Over the years, Dana was not only my partner in this endeavor but my mentor as well. This book is dedicated to the memory of my friend J. Dana Thompson.

Contents

Preface

More than 20 years ago, when I first visited the National Center for Atmospheric Research (NCAR) in Boulder, Colorado, as a postdoctoral fellow in the Advanced Study Program, I accidentally "stumbled" across El Niño. Here is how it happened. In 1972–73, a major El Niño event occurred off the coast of Peru. That event was linked directly to the collapse of the Peruvian fishing industry, which at that time had been the number 1 fishing industry in the world in terms of the total weight of fish catches. The collapse of the fishery generated considerable scientific interest in the biological impacts of El Niño and, more broadly, in the El Niño phenomenon.

The public attention that was generated by the negative consequences of the El Niño in 1972–73 was short lived, after which societal interest in the phenomenon remained dormant for some years. Only a few governments continued to encourage their scientists to understand the phenomenon better and to determine with greater accuracy how it might directly affect their economies and to forecast when the next such event might occur. The sharpest increase in public awareness about how El Niño events can impact on human activities came in the wake of the 1982–83 event. This has been touted by scientists as the biggest El Niño event in a century (biggest can be determined by a variety of factors: the sea surface temperature increase was larger than expected, warmer surface water spanned a larger portion of the Pacific Ocean's surface than in past events, and the impacts on ecosystems and societies were more devastating than during earlier El Niño episodes). The records show that there have been no other events equally as "big" in more than 400 years. This particular event, however, seemed to have captured the attention of the media, especially in North America, as a result of numerous climate-related problems around the world being blamed directly on El Niño influences.

One could effectively argue that El Niño started to become a household word, more or less, in February 1984, when *National Geographic* magazine chose to present to millions of its readers around the globe a photo essay about the 1982–83 event and its worldwide consequences. El Niño was even featured on the cover. Later that year, an El Niño article appeared in *Readers' Digest*, whose subscribers number in the tens of millions and whose editions appear worldwide in 18 languages.

Only toward the end of the most recent El Niño event(s) in the 1991–95 period did government agencies in various countries begin to show a much more serious concern about the impacts and applications side of the El Niño phenomenon. Before that time, it was up to the research interests of individual social scientists to "bother" themselves with undertaking research on the societal aspects of El Niño, often unsupported by outside funding. Australia is perhaps *the* leading example of a country that was ready to take El Niño events seriously. Although there had been some research and interest in El Niño events in that country's scientific circles, it was only after the 1982–83 event that Australians sought to look more closely at its impacts on various sectors of their economy, including public safety. In the social sciences, unlike what has happened in the last couple of decades within the physical and biological sciences, no network of researchers with a direct research interest in the social and economic aspects of El Niño events has formed. Yet, El Niño is a natural phenomenon, and improved information about it could yield great benefits to those who choose to use it judiciously. It can lead to more effective decision-making.

Michael Glantz
Boulder, Colorado
April 1996

Acknowledgments

Many people have helped me with various aspects of the El Niño story. I want to thank them for their assistance and patience in responding to my numerous pleas: Peter Gent, Mark Cane, Joseph Tribbia, Antonio Busalacchi, Wang Shao-wu, Tesfaye Haile, Kenneth Mooney, Gary Sharp, Warren Wooster, William Kellogg, George Kiladis, Roger Pielke Jr, Rick Katz, Stephen Zebiak, Claudia Nierenberg, Antonio Magalhães, Dale Jamieson, Elm Sturkol, Carl Hunt, Vicki Holzhauer, Maria Krenz, Brad McLain, Leslie Forehand, and Jan Hopper. Their critical reviews were extremely useful. Special thanks go to Michele Betsill for research and administrative support. The lion's share of thanks goes to D. Jan Stewart, who had the onerous task of preparing numerous drafts of the manuscript and editing the text. She exhibited great patience and perseverance while undertaking this task on top of all her other responsibilities, including teaching daring souls how to skydive. I am very much indebted to my Cambridge University Press editor, Tracey Sanderson, who provided me with sorely needed guidance and direction at important junctures in the preparation of this manuscript. My wife, Karen, has been an extremely understanding companion, having given up the summer of 1995 to El Niño! I owe her much more than the summer of 1996.

The National Center for Atmospheric Research is sponsored by the National Science Foundation.

1 Introduction

Climate is what you expect.
Weather is what you get.
Anon.

Weather and climate variability

Every year there are extreme climate-related problems around the globe, with droughts occurring in some places and floods in others. For example, the summer of 1988 witnessed a severe drought in the agricultural heartland of North America and extremely low streamflow in the basin of the mighty Mississippi River. Just a few years later, in the summer of 1993, a period of very heavy rains led to major flooding along the Upper Mississippi and Lower Missouri rivers and many of their tributaries in the United States Midwest. In the early 1990s, newspaper headlines noted that drought-related food shortages in southern Africa put about 80 million Africans at risk of famine. In early 1995, extreme flooding occurred in western Europe, shaking the confidence of countries such as the Netherlands in their ability to prevent natural catastrophes, and challenging their false belief that scientific and technological developments had buffered their societies from the consequences of such periods of extremely heavy rainfall. This was not unlike the situation in the 1970s and 1980s, when Canadian officials sought to "drought-proof" the climate-sensitive agricultural areas in the Canadian prairie provinces, only to realize the impossibility of such a daunting task.

The point is that record-setting climate events are occurring somewhere in the world each year. In fact, Sir John Houghton, head of a major international program designed to assess the level of present understanding of the science of climate change, has suggested that "records are being set every year and if there were a year without such an occurrence, that in itself would be record-setting." (J. T. Houghton, cited by Greenpeace International, 1994).

Nevertheless, it seems that in some years there are many more extreme meteorological events and resulting societal problems, such as droughts, floods, frosts, or blizzards, than one might expect, even if they were not record-setting. One such period was 1972–73, when severe droughts occurred in widely dispersed locations such as Australia and Indonesia, Brazil and Central America, India and in parts of sub-Saharan Africa, and heavy flooding occurred in Kenya, southern Brazil, and parts of Ecuador and Peru. At that time it was suggested that some of these widely dispersed climatic extremes might have had a common geographic origin – changes in sea surface temperatures in the Pacific Ocean (El Niño or EN) and changes in sea level atmospheric pressure across the Pacific basin (the Southern Oscillation or SO). These combined changes have come to be commonly referred to as El Niño events in the popular media and as ENSO (El Niño–Southern Oscillation) events in much of the scientific literature.

Very briefly, an El Niño event can be described as the appearance from time to time of warm sea surface water in the central and eastern Pacific Ocean near the equator. Folklore suggests that the term "El Niño" (literally, the Spanish phrase for "the Christ Child" or "Baby Jesus") was used by Peruvian fishermen who had noticed the annual appearance of warm water along the western coast by December of each year. In some years, the warming along the coast did not dissipate within the usual few months but lingered for more than a year. This too was called "El Niño." "El Niño" has now been broadened to include all kinds of sea surface warming in the equatorial Pacific. Scientists believe that El Niño events are related to anomalous weather extremes around the globe.

During the past couple of decades the public has learned of El Niño and its impacts in a sporadic way. It would be mentioned in the popular media only when a big El Niño event was believed to be under way. Many of those articles or news releases were simply reports of the events of the day and were devoid of in-depth discussion of the phenomenon. Once the El Niño event (or threat of it) had passed, the media's interest in it waned rapidly. One of the key reasons for undertaking the preparation of this book is to provide a user-friendly account of what El Niño is, what it does and why we, as members of different societies, need to have more than a passing, intermittent interest in it, limited for the most part to when it occurs every few years or so.

(El Niño and worldwide climate|

The associations between El Niño events and unusual changes in normal climate patterns (called anomalies) around the globe have been referred to by scientists as "teleconnections." These are known, as well as alleged, connections between El Niño events and changes in distant weather or climate-related processes. For example, there appears to be an association between El Niño events and droughts in various parts of the globe:

northern Australia, southeastern Africa, Northeast Brazil, parts of India, Central America, and so forth. There also appear to be linkages between El Niño events and a reduced number of tropical hurricanes occurring in a given year along the east coast of the USA as well as in the locations of tropical cyclones off the east coast of Australia, where they tend to shift equatorward by several hundred kilometers.

One location where ecosystems and human activities are known to be directly and, for the most part, adversely affected by El Niño is the area along the western coast of South America, specifically Peru, Ecuador, and northern Chile. Just about every event, regardless of whether weak or strong, has an impact on this region.

El Niño and societal impacts

El Niño is a natural phenomenon that recurs every few years. To varying degrees, it affects a large portion of the world's population. Potentially useful scientific information about El Niño and its impacts on society will probably go unused, unless there are sustained efforts to educate the public about how to realize the value of seemingly abstract scientific research findings. For this reason alone, it is important for the general public, for managers in various economic sectors, and for policymakers to know more about the El Niño phenomenon, including its teleconnections, and its implications for ecosystems and societies around the globe. The scientific literature and popular media are full of statements about the value to society of being forewarned about the possible onset of an El Niño. On an idealized, abstract level, it is not difficult to find value in forecasts of El Niño events or, for that matter, forecasts of any climate-related environmental change. However, when it comes to a specific El Niño event and its specific impacts in local areas worldwide, it becomes a highly speculative endeavor to place a precise value on such forecasts.

El Niño and Peru

The value of knowing more about El Niño to various sectors of Peruvian society has been mentioned in general statements since at least the end of the 1800s. For example, at an International Geographical Congress held in Lima, Peru, in the early 1890s, Peruvian geographer Federico Alfonzo Pezet stated that

> the existence of this counter-current [El Niño] is a known fact, and what is now wanted is that proper and definite studies, surveys, and observations should be undertaken in order to get to the bottom of the question, and

3

find out everything relating to this counter-current, and to the influence
which it appears to exercise in the regions where its action is most felt.
(Pezet, 1895, p. 605)

One of the major regional influences Pezet referred to was heavy rainfall
that extended well beyond a single season in northern Peru that usually
accompanied El Niño events.

One could easily argue that, at the end of the nineteenth century, El Niño
was of interest mainly to local populations along the western coast of South
America because of the associated disruptions in both normal (i.e.,
expected) rainfall patterns and reproductive and behavioral patterns of fish
and bird populations along the coast. Actually, concern was not so much
directed toward the apparent adverse impacts on fish as it was toward the
bird populations that lived off them.

During major El Niño events, fish populations, especially anchoveta,
would be reduced as a result of decreased food supply and would shift their
location, becoming less accessible to fish-eating birds. As a direct result,
millions of adult sea birds and their chicks would perish. The carcasses of
thousands of dead birds would wash up onto Peruvian beaches. In many
countries, the occasional high level of mortality among bird populations
might receive brief notoriety; not so for the sea birds of Peru. Various sea
birds earned the name of guano birds because they were highly valued in
Peru for their excrement (called guano). Guano was "discovered" by
European chemists in the early 1800s to be rich in nitrogen and phos-
phorus. It was to be used as an excellent fertilizer for agricultural fields in
Britain, Europe, and the USA. Guano was considered a valuable export
commodity for Peru between 1840 and 1880. El Niño-related reductions in
the guano bird population led to reduced production of guano in Peru's
guano-bird rookeries on the rocky Chincha Islands and coastal areas
(Figure 1.1).

El Niño and the world

Because of the advent of manufactured fertilizers and other trade-related
factors, Peru's ability to export guano waned and, as a result, birds and
guano production were no longer major generators of Peruvian concern
about El Niño or its ecological impacts. Interest in El Niño shifted in the
1950s to the exploitation of the anchoveta fish population for the purpose of
fishmeal production. With the collapse of that fishery in the mid-1970s, the
health of fish populations in the eastern equatorial Pacific no longer
generated global interest in El Niño episodes. Primary interest today
centers on the realization that El Niño is a Pacific basin-wide

Figure 1.1. The rocky islands along the coast of Peru provide nesting sites for "guano" birds. In the absence of an El Niño event, the birds are highly productive, consuming large quantities of fish, primarily anchoveta, that dwell near the ocean's surface. This is converted into guano, bird droppings that are used as a fertilizer for agriculture. When the warm waters appear in the region, signaling an El Niño (warm) event, the fish become fewer in number and are dispersed, becoming inaccessible to the birds. (Jaime Jahncke – IMARPE.)

phenomenon with regional perturbations of climate- and weather-related processes throughout most of the world.

Increasingly, the task of understanding El Niño is seen by climatologists and meteorologists as an important key to unlocking mysteries about tropical climate and weather patterns and, to a varying extent, their impacts on regions outside the tropics (called the extra-tropics). In fact, an increasing number of El Niño researchers now claim that they can reliably forecast the onset of El Niño. Depending on the particular researcher, claims for lead time (i.e., advanced warning) range from four to 12 months. Some of these claims are considered to be realistic and have, in turn, captured the attention of policymakers who continue to support physical science research on El Niño and forecasting efforts in a major way. Disruptions of regional climate patterns and of human activities during El Niño events reinforce the need for development of reliable, long-range, climate-related forecasts, which can then be used to reduce the impacts on society and on vulnerable ecosystems of climate and weather extremes.

El Niño and international science

Scientific research interest in El Niño has blossomed. El Niño research is no longer left to Peruvian scientists interested in sea birds or fish to investigate the phenomenon only for its adverse, local ecological consequences. Now, they have been joined by a small army of scientific researchers, drawn from all continents and from several academic disciplines, who are actively engaged in individual as well as collaborative research on El Niño-related topics. Their shared expectation is to resolve lingering mysteries about the phenomenon and to uncover the underlying mechanisms that perpetuate El Niño events and govern their life cycles. Such discoveries would probably enable scientists to forecast El Niño with a high degree of reliability several months to a year in advance of the onset, growth, and decay phases of the phenomenon.

Today, policymakers who have been funding science are asking questions about the value to society of its work. With limited national budgets, researchers are having to consider the usability of their research findings. Forecasting El Niño and, more broadly, the forecasting of climate variability from one year to the next, has potential benefits to society. El Niño research can easily be used to demonstrate how the scientific community can produce "usable science." The scientific community has only recently come to realize the need for sustained efforts to educate the public, and especially policymakers, about the importance of this phenomenon. There has also been a sharp increase in policymaker interest in identifying the environmental and societal consequences of El Niño.

El Niño as a "living" thing

Like many other processes in nature, El Niño comes and goes again and again. Like other recurring phenomena in nature, such as seasonal vegetation cycles, mountain snowpack, glacial advance and retreat, and sand dune movement, El Niño events wax and wane over time in their responses to climatic fluctuations.

As with any attempt to discuss a system with many interacting components, identifying the best place to begin is often difficult. For example, how might one best describe the human body? Before discussing the parts of the human body and their functions, one must have an idea of how the body as a whole works. By analogy, for El Niño it is necessary to describe the phenomenon (i.e., the body) and then to describe its components and their various interactions. In using such an approach, however, it is difficult to avoid some repetition. This inconvenience aside, the reader will gain a better picture of El Niño and its impacts on weather and climate anomalies around the globe.

The word "complex"

The scientific community relies heavily on the use of the term "complex." Atmospheric processes are complex; so too are oceanic processes. El Niño is the result of complex interactions between the atmosphere and the ocean. In reality everything is complex, from the electron that circles the nucleus of an atom to the far reaches of the universe.

However, in addition to the acceptance of scientific complexity, the term "complex" has also been used for a variety of other reasons. For example, "complex" has been used as an adjective to suggest that an understanding of the phenomenon under investigation cannot be known in its entirety. It has also been used to suggest that it will take a long time (i.e., much money) to understand it completely. "Complex" has also been used by scientists as a caveat to note "buyer beware"; that the users of such information should treat it as imperfect information. In some instances, it has been used to suggest that you (the reader) cannot possibly understand all that the scientist could tell you about the phenomenon, so he/she won't bother to try. In sum, the notion of complexity can be used, on the one hand, to expose the limits to our depth of knowledge or, on the other hand, to hide our ignorance. In this book the term will be used sparingly, assuming that readers are well aware of just how complicated are various natural processes and interactions. How much of the science of El Niño does the non-expert *need* to know? How much detail can be left out or generalized in a description of processes and events, while the account still conveys an understanding that is correct, even if not complete? That is the challenge of those who seek to write about scientific issues for those of us who are not physical scientists.

Chapter overview

Chapter 2 presents definitions of El Niño and brings to light a main source of the confusion that surrounds the phenomenon. It attempts to provide a broad definition for El Niño. Chapter 3, entitled "A tale of two histories," presents a brief history of the emergence and spread of scientific interest in El Niño and in the Southern Oscillation.

The fourth chapter, "Biography of El Niño," describes various characteristics and processes associated with El Niño and the Southern Oscillation and Chapter 5 discusses the 1982–83 event, which has been considered the biggest in a century. Attention is drawn to the importance for scientific research of the 1972–73 event.

Chapter 6 considers the value, in theory and in practice, of "Forecasting El Niño," providing a few examples of forecast successes and failures. Also mentioned is the unexpected behavior of air–sea interactions in the equatorial Pacific in the 1991–95 period, as well as the need for an international institute for climate prediction on an interannual basis.

Many people are interested in the equatorial Pacific, insofar as they believe that their regional climate is affected by events there. Chapter 7 focuses on the linkages of weather or climate anomalies in distant locations, called teleconnections, believed to be associated with El Niño.

How researchers monitor, investigate and forecast El Niño events is briefly discussed in Chapter 8, and Chapter 9 outlines some of the major post-war international science programs, starting with the International Geophysical Year (1957–58) and extending into the beginning of the next century. This chapter also addresses questions raised about how global warming of the atmosphere might affect El Niño events. If the atmosphere were to warm by a few degrees Celsius in the last half of the next century, that would probably affect the El Niño process in as yet unknown ways.

Why care about El Niño? is addressed in Chapter 10, which suggests why people who are not directly affected by its effects should also take interest in the phenomenon. The specific examples provided suggest that there are costs associated with not using El Niño information that is already available and considered reliable.

The eleventh chapter is a collection of thoughts by researchers whose activities span several decades. These scientists represent various disciplines and activities and they were asked to provide a few paragraphs about an aspect of El Niño that he or she wished to share with the readers.

The final chapter addresses the contentious issue of "Usable Science," what it is and who decides which research findings are of direct use to society on a time scale of interest to present-day decisionmakers.

The crossword puzzle (Figure 1.2, overleaf) is based on the El Niño phenomenon. While most readers would have trouble at the outset answering the puzzle's clues, it is hoped that they will be able to attack it after having read *Currents of change: El Niño's impact on climate and society.*

ACROSS

1. One of the Southern Oscillation's northern sisters
3. Decade-long research program on Pacific air–sea interaction
6. Researcher associated with Niño 3
8. Highly productive oceanic regions
13. Statistical measure for identifying teleconnections
14. Concept used to manage fisheries
15. Fishmeal production by-product
16. Peru's fishing rival
20. 1970s biological experiment on coastal upwelling
21. Shares ecological niche with anchoveta
22. He identified the Southern Oscillation
26. Spanish word for "here"
28. Measure of equatorial Pacific pressure changes
30. Pressure pattern affecting North American weather
31. Major producer of fishmeal
32. Fishing net characteristic determining the size of fish that can be caught
33. Limits the movement of internal (Kelvin) waves
36. Prestigious US scientific academy
39. Synthetic fiber that revolutionized fishing in Peru
40. He identified the link between the EN and the SO
41. Word for metric ton

DOWN

2. Animal feed supplement source taken from the sea
3. Linkage between remote climate anomalies and El Niño
4. He created an historical time series of El Niño events
5. An alleged by-product of El Niño-related drought in Indonesia
7. El Niño's other name
9. University of Hawaii professor who measured sea level changes with tide gauges in the equatorial Pacific
10. The opposite of El Niño's cold event
11. Site of the first major workshop on El Niño in the mid-1970s
12. South-to-north circulation pattern in the Pacific
16. Site of Peru's Marine Institute
17. Rainfall failures in this country prompted search for teleconnections
18. Location used in the Southern Oscillation Index
19. Site of the Earth Summit in June 1992
21. Fish off Pacific Northwest coast affected by El Niño events
23. The Christ Child
24. Thermal reservoir in the western Pacific Ocean
25. Primary use of Peruvian anchovy
27. Spanish word for "what"
29. Peru's research center concerned with effects of El Niño
34. Fertilizer; bird excrement
35. Program created by four western South American countries to assess El Niño impacts
37. A second-choice animal feed if fishmeal is unavailable
38. A change in this ocean characteristic signals the onset of an El Niño (abbreviation)

Figure 1.2. Solving the El Niño mystery. Crossword puzzle by M. H. Glantz. All terms in the El Niño crossword puzzle are referred to in the text. The solution may be found on p. 188.

Section I
Emerging interest in El Niño

2 El Niño

El Niño definitions

The term "El Niño" means different things to different people. In Spanish, *el niño* means small boy or child. With capital letters, El Niño refers to Jesus as an infant. To Peruvians, it has an additional meaning: a particular intermittent warm ocean current that moves southward along its coast. They gave the ocean current the name El Niño at some time before the beginning of the twentieth century, although its exact origin and "birth date" remain unknown. The popular contemporary version of how it got its name relates to the fact that warm waters appear off the coast of Peru seasonally, beginning around Christmas time (i.e., during the Southern Hemisphere summer, which is the Northern Hemisphere winter), temporarily replacing the usually cold waters in that region for a few months. The normally cold waters along the coast are the result of coastal upwelling processes by which deep, cold, nutrient-rich water wells up to the ocean's sunlit surface (called the euphotic zone).

At a Geography Society meeting in Lima in 1892, Peruvian navy captain Camilo Carrillo was apparently the source of information (and rumor) about how the El Niño current got its name. He made the following statement, which has now been repeated many times:

> Peruvian sailors from the port of Paita in northern Peru, who frequently navigate along the coast in small crafts, either to the north or to the south of Paita, named this current "El Niño" without doubt because it is most noticeable and felt after Christmas.
>
> (Carrillo, 1892, p. 84)

Occasionally, the seasonally warmer water that appeared off the coast of Peru and Ecuador (a region referred to as the eastern equatorial Pacific) would linger longer than a few months, sometimes lasting well into the following year. These prolonged "invasions" (more correctly, appearances) of warm water have led to pronounced disruptions of regional coastal ecosystems and socioeconomic activities.

In his speech to the International Geographical Congress in Lima, Pezet

(1895, p. 605) noted: "that this hot current has caused the great rainfalls in the rainless regions of Peru appears a fact, as it has been observed that these heavy rains have taken place during the summers of excessive heat". Even though we do not know when El Niño events were recognized as such, we do know that they have occurred over millennia, as the impact of heavy rains and flooding has left its marks on the natural environment in Peru and Equador.

As of the beginning of the twentieth century, the connection between the occurrence of El Niño and the various changes in the natural environment around the tropics, from the east coast of the African continent to the west coast of South America, had not yet been made. El Niño's impacts were only of concern in Peru and Ecuador, where they were considered to be local manifestations of a local oceanic or atmospheric variation.

As early as the mid-1970s, El Niño had acquired several definitions (Barnett, 1977). By the mid-1990s, several dozen El Niño definitions could be found in scientific articles and books, ranging from simple to complex ones. The following two definitions serve as examples.

> El Niño: A 12–18 month period during which anomalously warm sea surface temperatures occur in the eastern half of the equatorial Pacific. Moderate or strong El Niño events occur irregularly, about once every 5–6 years or so on average.
>
> (Gray, 1993)

> Originally ... an El Niño referred to warm current flows along the coasts of Ecuador and Peru in January, February, and March, and the resulting impact on local weather. ... The second name, ENSO, more generally refers to events from the mid-Pacific to the South American coast, taking into account the irregular oscillation in pressure between the east and west Pacific.
>
> (Palca, 1986)

However, some common aspects of El Niño do recur in these definitions. El Niño

- is an anomalous warming of surface water,
- appears off the coasts of Ecuador and northern Peru (sometimes Chile),
- is linked to changes in pressure at sea level across the Pacific Ocean (Southern Oscillation),
- recurs but not at regular intervals,
- involves sea surface temperature increases in the eastern and central Pacific,
- is a warm southward-flowing current off the coast of Peru,
- accompanies a slackening of westward-flowing equatorial trade winds,

- returns around Christmas time,
- lasts between 12 and 18 months.

An example of a technical definition that has been used to identify an El Niño event includes the following elements:

> The sea surface temperature (SST) index represents the season SST anomaly (within 4° of the equator from 160°W to the South American coast). ... The anomaly had to be positive for at least three seasons, and be at least 0.5°C above the average for at least one season, while the SOI [Southern Oscillation Index based on the difference in pressure at sea level between Darwin (Australia) and Tahiti] *had to remain negative and below* −1.0 *for the same duration.*
>
> (Kiladis and van Loon, 1988)

Several definitions of El Niño also contain comments about what an El Niño event does: for example, comments such as "the reduced welling up to the surface of deep cold water," "the appearance of nutrient-poor water," "causes weather changes around most of the globe," and so forth.

Periods of cold sea surface temperatures are followed by periods of warm sea surface temperatures, which are usually followed (once again) by cold, or at least near-normal, sea surface temperatures, and so on. This recurring, variable pattern is part of the normal year-to-year climate variability and not apart from it. Suggestions, however, about its return period abound. e.g., 2–10 years, 4–7 years, 3–4 years, 3–7 years, 5–6 years, 5–7 years.

Almost all definitions refer to the *anomalous* warming of sea surface temperatures off the coast of Peru and in the central equatorial Pacific Ocean. Perhaps in a strict statistical sense, it is anomalous. They are changes from the seasonal average sea surface temperatures, departures which can be arbitrarily fixed at 1, 2, 3, or more degrees Celsius. In another sense, however, these changes from average conditions are to be expected. On one hand, an El Niño might be called an anomaly. From a different perspective, however, it could also be considered a normal phenomenon. If the scientific community that researches El Niño is showing signs of difficulty in keeping these definitions straight, can the public and policymakers be any less confused?

When people hear about El Niño, many have a general notion that "something" is happening in the waters "somewhere" off the coast of Peru. As scientists continue to discuss El Niño and the media write about it, the public increases its knowledge about the interactions between the ocean and the atmosphere. Much of the public now knows that environmental changes in the equatorial Pacific region are more far-reaching than just those that take place along the Peruvian coast. Therefore, the scientific literature is referring increasingly to the broader Pacific basin-wide changes in sea surface temperature and surface pressure oscillations as ENSO.

During the course of the past decade, the public has become familiar with and used to the term El Niño through its use in the media. In fact, media representatives have even stated that ENSO is a difficult concept to explain in simple terms to the general public. As a result, even the larger basin-wide changes in the equatorial Pacific are often still referred to by the public, the media, and many scientists as El Niño.

Dictionary definitions of common words often provide several meanings that have been attributed to one word. Sometimes those meanings conflict with each other. Take, for example, the word "leader." One definition notes that a leader has the *ability to lead*; another states that a leader must *take action*. While these two meanings may overlap, they do not convey the same message. By analogy, then, El Niño can have more than a single meaning. It can encompass both a localized coastal warming of sea surface temperatures *and* the much broader basin-wide ENSO event in the equatorial Pacific. It is interesting to note that some researchers have used those words – El Niño and ENSO – interchangeably even within the same publication. The definition of El Niño in the box below is offered, encompassing the range of meanings and terms attributed to it. For the reader's sake, "El Niño" is used in this book to describe both the local and regional-scale warming of the sea's surface.

El Niño \ 'el nē' nyō *noun* [spanish] \ 1: The Christ Child   2: the name given by Peruvian sailors to a seasonal, warm southward-moving current along the Peruvian coast ⟨la corriente del niño⟩   3: name given to the occasional return of unusually warm water in the normally cold water [upwelling] region along the Peruvian coast, disrupting local fish and bird populations   4: name given to a Pacific basin-wide increase in both sea surface temperatures in the central and/or eastern equatorial Pacific Ocean and in sea level atmospheric pressure in the western Pacific (Southern Oscillation)   5: used interchangeably with ENSO (El Niño–Southern Oscillation), which describes the basin-wide changes in air–sea interaction in the equatorial Pacific region   6: ENSO warm event *synonym* warm event   *antonym* SEE La Niña \ [Spanish] \ the young girl; cold event; ENSO cold event; non-El Niño year; anti-El Niño or anti-ENSO (pejorative); El Viejo \ 'el vyā hō \ noun [Spanish] \ the old man

M. H. Glantz

Measuring El Niño

There are several "types" of El Niño event. For example, they can vary in size. The measuring of an event's size can depend on quantitative indicators. One of the most obvious and important indicators is the increase in sea surface temperatures in either the central or eastern equatorial Pacific (Figure 2.1).

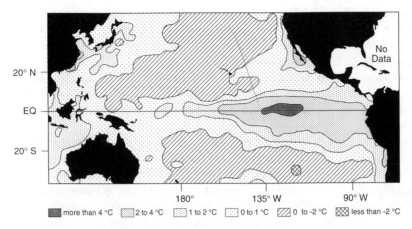

Figure 2.1. Observed sea surface temperature anomalies for December 1982–February 1983 in the Pacific. (From NOAA/OGP, 1992.)

The larger the increase in temperature above normal, the larger the event. Scientists often refer to the size of El Niño as very weak, weak, moderate, strong, or very strong. "It is assumed that the stronger the event, the greater the amount of damage, destruction, and cost to the nation" (Quinn et al., 1987, p. 14 453). According to Quinn et al.,

Very strong events show extreme amounts of rainfall, flood waters and destruction in Peru, and coastal sea surface temperatures usually reach values of more than 7 deg.C above normal during some months of the Southern Hemisphere summer and autumn seasons.

Strong events, in addition to showing large amounts of rainfall and coastal flooding and significant reports of destruction, exhibit coastal sea surface temperatures in the range 3–5 deg.C above normal during several months of the Southern Hemisphere summer and autumn seasons.

Moderate events, in addition to showing above-normal rainfall and coastal flooding and a lower level of destruction, generally show coastal sea surface temperatures in the range 2–3 deg.C above normal in the Southern Hemisphere summer and autumn seasons.

Other factors used to determine the "size" of an El Niño are the geographic location and the area covered by the anomalously warm pool of sea water. Generally speaking, the larger the area of the warm ocean surface, the bigger the event. A small event might be one where the warming is restricted to the western coast of South America, while in a larger one both the central and eastern equatorial Pacific would be encompassed by the warm water anomaly.

Yet another way to measure the size of an El Niño is by how long it lasts. Scientists suggest that El Niño events usually last for 12 to 18 months. Major ones rarely last longer than a few years; however, longer events have been witnessed a few times this century. The latest prolonged El Niño began in 1991. According to the United States National Weather Service (Halpert *et al.*, 1994), the event that began in 1991 was not the only unusual one this century. El Niño events of similar duration occurred in 1911–13 and in 1939–41.

In 1991 an El Niño event began that some researchers have suggested spanned at least three calendar years, 1991–93. Others have suggested that it did not end in 1993 but continued into early 1995. Was it one long event, a few smaller ones? While the 1982–83 event was considered by many to have been the most devastating in a century, the 1991 event may eventually be shown to have been among the longest. Perhaps another category, "*extraordinary*," should be added to describe the size of El Niño.

Determining an El Niño's size through its impacts

Determining the size of an El Niño can also involve subjective elements, often presented as scientific facts. In the late 1970s, William Quinn and colleagues identified and categorized El Niño events going as far back in history as the early 1500s. They used direct observations of recent El Niño events and their direct impacts in the twentieth century. They relied on proxy information to help them to reconstruct El Niño events in historical time. Proxy information is an indirect indicator of El Niño, where these events cannot be directly monitored. Researchers have gathered such indirect information on rainfall and ocean temperatures from a variety of sources, including personal diaries of travelers in the region, guano-mining records, plantation records in Indonesia, ships' logs, and physical and historical evidence of floods and mudslides that took place centuries ago (Quinn *et al.*, 1987). Examples of other indirect observations of El Niño are as follows: El Niño events are often accompanied by heavy rainfall, an increase in the number of warm-water fish species in the upwelling region, and by the appearance of thousands of dead guano birds along the usually arid coast of Peru. These factors are also used to determine the size of a particular El Niño event.

In addition to examining the physical characteristics of an El Niño, one can consider El Niño's impacts on human societies. Some El Niño events have had only local or regional impacts on human activities through, for example, heavy rains, flooding, and mudslides that have destroyed villages and the transportation infrastructure. Larger events can, for example, have major negative effects, such as those on food production in regions that are located at great distances from the equatorial Pacific, e.g., southern Africa or northeastern Brazil.

El Niño's impact on ecosystems also indicates its size. For example, M. Coffroth and colleagues (cited by Glynn, 1990) observed that "1982–83 witnessed the most widespread coral bleaching and mortality in recorded history" (p. 141). Worldwide ecological changes coincided with this "event of the century" (Hansen, 1990, p. 1). Thus, the larger the El Niño, the more likely there will be both an increase in the number of remote locations that are affected by it and an increase in the total cost of all its adverse impacts. Figure 2.2 provides examples of El Niño's impacts on regional climate, ecosystems, and on human activities.

El Niño as a natural hazard

There is a long-standing community of researchers with a focus on natural hazards, for the most part rapid-onset events, such as river flooding, blizzards, avalanches, tsunamis, earthquakes, and hurricanes. Although it meets many of the criteria used to describe them, El Niño has not as yet made this list of such hazards. Ian Burton and colleagues (1993, pp. 35–6) have listed characteristics that define a hazardous event: magnitude, frequency, duration, areal extent, speed of onset, spatial dispersion, and temporal spacing, each of which they define as follows:

- *Magnitude*: only those occurrences that exceed some common level of magnitude are extreme.
- *Frequency*: how often an event of a given magnitude may be expected to occur in the long-run average.
- *Duration*: the length of time over which a hazardous event persists, the onset to peak period.
- *Areal extent*: the space covered by the hazardous event.
- *Speed of onset*: the length of time between the first appearance of an event and its peak.
- *Spatial dispersion*: the pattern of distribution over the space in which its [impacts] can occur.
- *Temporal spacing*: the sequencing of events, ranging along a continuum from random to periodic.

These characteristics apply well to El Niño. The *magnitude* of an El Niño is defined by degree of departure from a long-term average of anomalously warm sea surface temperatures in the central and eastern Pacific. *Frequency* relates to its return period, which scientists have suggested is of the order of 2 to 10 years (more specifically, one could argue that a major El Niño occurs every 8 to 11 years, and a minor one every 2 or 3 years). The *duration* of El Niño events is 12 to 18 months, with a few notable exceptions. The *areal*

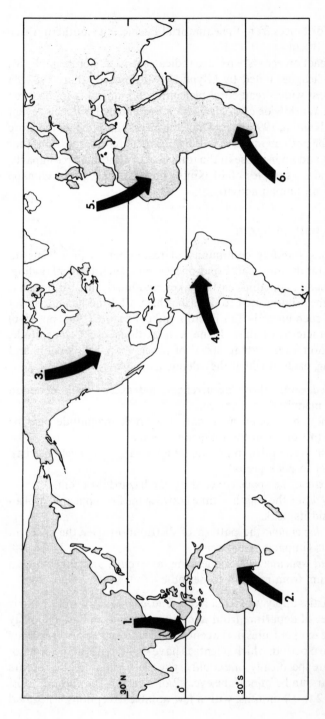

Figure 2.2. World map with description of some major climate impacts related to the 1982–83 El Niño event. (From Glantz, 1984.)

Key to Figure 2.2

Arrow 1. Indonesia was plagued with severe drought, resulting in reduced agricultural output, especially rice, and in famine, malnutrition, disease, and hundreds of deaths. This drought came at a bad time, as this country had made great strides toward self-sufficiency in food production. In the few years immediately preceding the 1982–83 El Niño, it was emerging as a rice exporter. This drought, however, coupled with worldwide recession, huge foreign debts, and declining oil revenues, has set back Indonesia's economic development goals for the near term.

Arrow 2. Australia had its worst drought this century. Agricultural and livestock losses, along with widespread bushfires mainly in the southeastern part of the country, resulted in billions of dollars in lost revenues. An Australian journalist wrote that "the drought is not just a rural catastrophe, it is a national disaster." The drought has been linked to El Niño.

Arrow 3. The eastern part of the United States was favorably affected by its warmest winter in 25 years. According to an estimate by the National Oceanic and Atmospheric Administration, energy savings were on the order of $US500 million. (The opposite was the case, however, during the cold winter that accompanied the 1976–77 El Niño.) The United States once again was adversely affected by devastating coastal storms and mudslides along the California coast, flooding in the southern states, and drought in the north central states, reducing corn and soybean production. Salmon harvest along the Pacific northwest coast was sharply reduced.

Arrow 4. In addition to the highly publicized damage to infrastructure and agriculture in Peru and Ecuador as a result of heavy flooding during this El Niño, there were severe droughts in southern Peru and Bolivia. A major drought continued in northeast Brazil, adversely affecting food production, human health, and the environment, and prompted migration out of the region into already crowded cities along the coast and to the south. There also were destructive floods in southern Brazil, northern Argentina, and Paraguay.

Arrow 5. Large expanses in Africa have been affected by drought. For example, the West African Sahel, once again, has been plagued by a major drought. While the human and livestock deaths resulting from this drought appear to be fewer than those in 1972–73, the food production situation is considered poor. The view that the Sahel has been in the midst of a long-term trend of below-average rainfall since 1968 is gaining credibility.

Arrow 6. Southern Africa has witnessed some of its worst droughts this century. This was not the case during the 1972–73 event. In 1983, for example, the Republic of South Africa, a major grain producer in the region, was forced to import from the United States about 1.5 million tonnes of corn to replace what was lost in their drought. Zimbabwe, a regional supplier of food, also was devastated by drought and was forced to appeal for food assistance from the international community. Likewise, Botswana, Mozambique, Angola, Lesotho, and Zambia, and the so-called Black National Homelands in the Republic of South Africa had their economies devastated by the drought of 1982–83.

extent could be interpreted to mean the spatial extent around the globe of the impacts of El Niño and its teleconnections. This would vary directly with the severity of the event, with major El Niño events being linked to major worldwide impacts and minor ones linked to localized or regional impacts. The *speed of onset* of El Niño is of the order of months. Occasionally, however, events have begun, only to collapse after a few months. *Spatial dispersion* refers to the area in the central and eastern Pacific that is encompassed by the anomalously warm sea surface temperatures. *Temporal spacing*, with respect to El Niño, refers to the return period which, on average, is 4.5 years.

Because the characteristics of an El Niño clearly meet the criteria used to define a natural hazard, El Niño merits inclusion in the list of natural hazards. An explicit designation could help to improve the level of research on its societal aspects, as has been the case with other designated natural hazards.

A note of caution

Societies change over time in a variety of ways. For example, populations are increasing. There are also shifts in the locations of where greater numbers of people live. The societal use of new technologies and techniques (i.e., ways of doing things) can alter the risks associated with the impacts of natural hazards. Changes that occur in society, many of which are the result of human decisions, can alter the degree of vulnerability of societies (often for the worse) in the face of climate-related hazards such as El Niño.

An example of this could be the population shift over the past several decades in the USA toward the Atlantic and Gulf coasts. A hurricane that makes landfall along the United States coastline today – having the same intensity, magnitude, and duration as that of a similar one a few decades ago – would probably cause much greater damage to life and property, because of the increases in population density in human coastal settlements and the infrastructures on which they depend. So, when seeking to measure the intensity of an El Niño event, it can be very misleading to focus on, and blame, El Niño for the "damage" that it leaves in its wake. Part of that damage (or, stated in another way, the increase in the risk of damage) can be directly attributable both to government and to private decisionmakers at various levels who permitted, if not encouraged, the demographic shift toward these coastal areas. Essentially, they were encouraging an increasing number of people to move into harm's way. Therefore, the severity of societal impacts attributed to El Niño events must be assessed with care before one attributes those impacts to an El Niño that happened to take place at the same time as unrelated variations in demographics at the local or regional level.

3 A tale of two histories

For much of the twentieth century El Niño and the Southern Oscillation have been studied as separate processes. Some researchers were interested in El Niño events in the equatorial Pacific Ocean and others focused on the Southern Oscillation occurring in the atmosphere from the Indian to the Pacific Ocean. It was not until the mid-1960s that Jacob Bjerknes, atmospheric scientist at the University of California at Los Angeles, voiced his ideas about the physical mechanisms linking these two seemingly separate processes. To understand El Niño in a broader context, it would be useful to understand the history of the development of interest in El Niño and in the Southern Oscillation.

History 1: Interest in El Niño

Interest in the El Niño phenomenon, as such, goes back at least to the middle decades of the 1800s. At that time, its adverse effects on guano birds (i.e., sea birds such as cormorants, gannets, and pelicans) and guano production in Peru had already been observed. Guano was "mined" throughout the rest of the century, despite the European decline in demand for it. By 1900, Peruvian authorities realized that guano was, in essence, being mined at an alarming and unsustainable rate. As a result, in the first decade of the twentieth century, the Peruvian government established a Guano Administration Company to oversee and control the digging up of guano that had been deposited over millennia along Peru's rocky coast and on its offshore uninhabited rock islands (Figures 3.1 and 3.2).

Guano birds along the Peruvian coast live off fish populations that dwell near the ocean's surface (these are called pelagic fish), primarily the anchoveta. The conditions in Peru's coastal waters are usually optimal for anchoveta populations. However, changes in physical, biological, and social conditions do occur along the Peruvian coast and they can be devastating to sea bird populations. Some of these conditions, which are occasionally perturbed by El Niño, are described briefly in the following paragraphs.

Figure 3.1. A gunny sack army breaks, bags and totes guano on Las Viejas Island. Because the slope is too steep for rails, local Indians carry the fertilizer to a seaside loading platform. A sack draped over the head cushions the burden and prevents it slipping. Centuries ago these men's ancestors performed a similar task under the Incas. (G. de Reparaz/National Geographic Image Collection.)

Physical setting

The rotation of the Earth, combined with the winds that tend to blow equatorward and offshore along the west coast of South America, pushes coastal surface water toward the open ocean and away from the land. As a result, cold water is drawn up from the ocean's depths to replace the warmer displaced surface water. This process is referred to as coastal upwelling.

Figure 3.2. Birds built a mountain of guano on the Central Chincha Islands in two and a half millennia; men carted it away in a few years. An orgy of exploitation in the 1860s stripped Chincha of its valuable cover. Measured against the workers, the heap is nearly 20 meters high. In some places the guano rose twice that much above bedrock; earlier digging reduced this pile. This old photograph shows Chinese laborers at work. (C. S. Merriman/National Geographic Image Collection.)

Coastal upwelling processes create regions in the ocean that are biologically highly productive (Figure 3.3).

The upwelling of ocean water brings chemicals into the sunlit layer of the ocean. They are converted to nutrients through photosynthesis for phytoplankton at the bottom of the food chain. The plants are eaten by zooplankton and fish populations, many of which are then consumed by guano birds. As one scientist noted in the first decade of this century, the anchoveta "are preyed upon by the flocks of cormorants, pelicans and gannets, and other abundant sea birds. ... The anchoveta, then, is not only ... the food of the larger fishes, but, as the food of the birds, it is the source

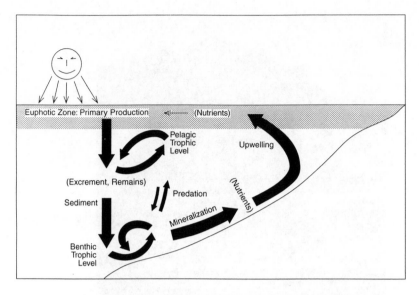

Figure 3.3. Nutrient cycling in an idealized coastal upwelling system.

from which is derived each year probably a score of thousands of tons of high-grade bird guano" (R. E. Coker, 1908, cited by Murphy, 1923).

The coastal upwelling phenomenon usually occurs along the western coasts of continents (with the exception of coastal upwelling along Somalia's coast in northeast Africa) in both the Northern and Southern Hemispheres.

Coastal upwelling

Coastal upwelling regions from around the globe make up about 0.1% of the ocean's surface area but provide more than 40% of all the commercial fish captured globally. The cold water, upwelling along the coast, tends to suppress rain-producing processes in the atmosphere and, as a result, upwelling regions are usually found next to coastal deserts. Coastal upwelling regions are shown in Figure 3.4.

Biological setting

Biological productivity is measured in terms of the rate of fixation of carbon by photosynthesis. The productivity of an upwelling ecosystem is measured in part by the amount of nutrients that is brought into the sunlit layer near the surface. According to David Cushing,

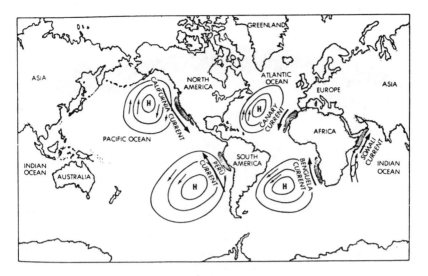

Figure 3.4. Major coastal upwelling regions of the world and the sea level atmospheric pressure systems that influence them.

Each [*upwelling region*] moves poleward as spring gives way to summer and each is two or three hundred kilometres broad in biological terms, even if the prominent physical processes are confined to a band within about 50 km of the coast.

(Cushing, 1982, p. 19)

Of these highly productive regions, Peru's is considered to be one of the best in terms of tonnage of fish landed (mostly anchoveta) (Figure 3.5). Before 1960, Peru was not noted for its fishing. However, by the mid-1960s and early 1970s, it had become the world's number 1 fishing nation.

When El Niño events occur, coastal upwelling processes are altered to such an extent that fish behavior within and among species becomes modified in major ways. Anchoveta, for example, disperse and dwell deeper in the ocean. Patterns of reproduction and migration change for different fish species, with some reproducing less in the temporarily altered marine environment. Some fish populations such as sardines fare well in the new but temporary warm surface water environment. More specific to Peruvian interests, the standing stock of anchoveta (that is, the total population from which future generations are to be produced) becomes reduced for a variety of reasons, including higher mortality and lower fecundity. As a result of changes in the behavior of the anchoveta population, the fish become much less accessible to the guano birds, causing starvation and death of hundreds of thousands or even millions of birds, depending on the magnitude and

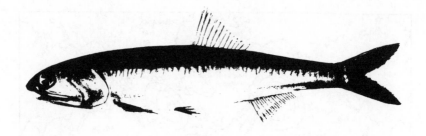

Figure 3.5. Anchoveta (Engraulis ringens Jenyns). Adult, actual size 17 cm. (Art courtesy of the Instituto del Mar del Peru.)

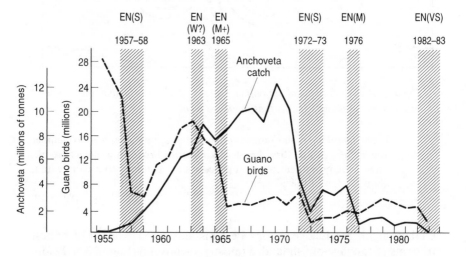

Figure 3.6. Fluctuation of the anchoveta catch and guano bird populations (cormorant, gannet, and brown pelican) off the coast of Peru in relation to El Niño events. W, weak; M, moderate; S, strong; VS, very strong. (From Jordán, 1991.)

intensity of the particular El Niño episode. Figure 3.6 provides an example of the adverse impacts over the long term on the various Peruvian guano birds of the combination of El Niño events and heavy commercial fishing pressures.

Societal setting

The Guano Administration Company and its main supporters among the Peruvian agricultural elite managed for many decades to block the development of large-scale commercial anchoveta fishing ventures. Apparently, they were able to argue successfully within the nation's highest

Figure 3.7. Peruvian modern fishmeal processing plant along the barren Peruvian coast about two hours south of Lima.

political circles that there were not enough fish in Peru's coastal waters to sustain both a viable guano-mining industry *and* a viable anchoveta fishing sector. These predators – birds and fishermen – would be competing for the same resources in order to maintain their livelihoods. It is important to note that the Peruvian anchoveta were not caught to be consumed directly by humans. They were to be processed into fishmeal (Figure 3.7) for use as a feed supplement for export primarily at that time to the rapidly expanding North American poultry industry and as fish oil for domestic use.

In the early 1950s, however, entrepreneurs interested in developing a Peruvian commercial fishery had convinced politicians to allow them to establish a commercial fishing industry, winning out over those who opposed its development. The arguments of the Peruvian investors in favor of establishing a national fishery received a major boost when the California sardine fishery collapsed from overfishing, sparking an increase in the demand for anchoveta fishmeal for export. The lure of being able to sell fishmeal for hard currency (i.e., US dollars) in the international marketplace put great pressure on Peruvian policymakers to open up Peru's coastal waters for exploitation by a nascent commercial fishing fleet. In fact, in 1953, the first fishmeal-processing factory had been brought into Peru from a defunct California sardine fishery. It was clandestinely reassembled at a remote site along Peru's arid coast without government permission. Figure 3.8 provides an overview of the fishery.

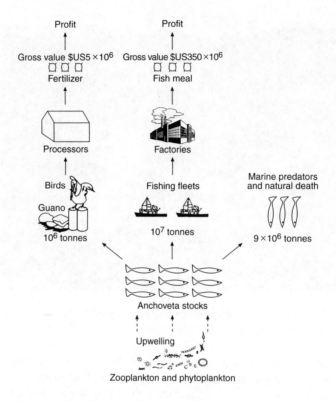

Figure 3.8. A diagram showing biological, economic, and social components involved in human exploitation of the natural fertility of the Peru Current as of the early 1970s. (From Paulik, 1981. Copyright © 1981 John Wiley & Sons, Inc. Reprinted by permission of John Wiley & Sons Inc.)

Robert Cushman Murphy, a zoologist for the American Museum of Natural History in New York City, studied guano birds along the Peruvian coast for several decades. He noted in a 1954 report for the Guano Administration Company to the Peruvian government that the development of an anchoveta fishery would, without doubt, lead to the demise of the guano bird population. He based his argument on the view that these birds consumed only the number of fish that they needed to satisfy their immediate food needs, whereas fishermen were insatiable predators. Fishermen would take from the ocean as many fish as their nets could capture and their boats could hold, and that they could sell to processing factories (Murphy, 1954). From the official opening of the anchoveta fishery in the early 1950s, a doubling of the officially reported anchoveta catch was repeated every year over the preceding year from the mid to late 1950s. Very likely, many landings went unreported.

Although a major El Niño took place in 1957–58, there was little, if any, direct reference to it in the popular press outside of Peru's borders. The most obvious reason that there were no visible effects on the productivity of the Peruvian anchoveta fishery was that by the late 1950s the anchoveta fishery had not as yet grown very large. Fish landings at the time were apparently still well below what scientists considered to be the maximum sustainable yield (MSY), i.e., the yield of fish that could be sustained over an indefinite period of time. Hence, those entrepreneurs and fishermen in the newly emerging Peruvian fishing industry showed little concern, if any, for El Niño or for its possible impacts on the anchoveta fish population that it was exploiting.

With the occurrence of the 1957–58 El Niño, the term's usage was deliberately broadened by the scientific community. It was viewed as a phenomenon common to other major coastal upwelling regions, especially the one off the coast of California. As early as 1959, oceanographer Warren Wooster commented on the concern that had been raised at that time about the use and misuse of the term El Niño. According to Wooster,

> Some readers ... have objected to the use of the name *El Niño* to identify the general phenomenon, feeling that previous usage restricts the term to the Peruvian Coast. If a more appropriate generic term can be found, I would recommend its use. I have used *El Niño* in a broad sense to emphasize that the Peruvian *Niño* is *not* a unique phenomenon, but is rather merely a striking example of a wide-spread occurrence.
>
> (Wooster, 1959, p. 45)

By the mid-1960s, marine biologists at the then recently established Peruvian Marine Institute (IMARPE) had become concerned about the increasing and uncontrolled levels of fishing due to pressures from fishing boat owners, from banks that had loaned funds for boards and equipment, from fishermen, and from the demands for raw material by the highly competitive fishmeal factories. They also realized that El Niño events, combined with increasing fishing pressure, could weaken the basic population of fish from which future populations were supposed to come. While the El Niño event that occurred in 1965 appeared to have reduced anchoveta fish landings only slightly, it had a devastating impact on the guano bird population, an impact from which it has yet to recover. It served as a "wake-up call" to some elements of the fishing industry and members of the Peruvian government, alerting them to the potential problems about which marine biologists were already aware. From the perspective of the anchoveta, El Niño events are in some ways like other predators – the guano birds and the fishing vessels – taking a share of their population. With fishing pressures on the anchoveta numbers mounting, policymakers and fishery managers were coming to realize that the anchoveta were not a limitless resource.

In response to those emerging concerns, biologists calculated that the maximum sustainable yield of the anchoveta population was about 9.5 million tonnes per year. They calculated that the guano birds required about 2 million tonnes, with a further 7.5 million tonnes going to the fishermen. Some American fisheries scientists, as well as some Peruvians, actually called for the deliberate destruction of the remaining guano-bird population in order to "free up" about two million additional tonnes of anchoveta for Peruvian fishmeal-processing plants. Fortunately, this suggestion was not acted upon.

In 1968, the Peruvian military overthrew the government. Up to the time of the *coup d'état*, the fishing sector had been under the administrative jurisdiction of the Ministry of Agriculture. This Ministry, as the name implies, put most of its attention on the agricultural sector, neglecting the management needs of the fishing sector. Successive Peruvian governments up to the time of the 1968 coup appeared to have had little awareness of just how important the fishing sector was to Peru's economy. The fishing sector, in its heyday in the late 1960s, was responsible for generating almost one-third of Peru's foreign exchange earnings. Hard currency earnings such as the US dollar, British pound, French franc, or German mark are extremely valuable to developing nations, making it possible for Peruvians to buy new technologies and other foreign goods.

From low-level bureaucrats to high-level Peruvian policymakers, all of whom were unfamiliar with either fish biology or the already known consequences of the overexploitation of a fish population, the coastal ocean appeared to be an endless source of fish and, therefore, of foreign exchange. Perceptions such as these caused Peruvian policymakers to disregard the advice of their own national scientists about the urgent need to tighten up the management of the fishery in order to save the resource itself. To do otherwise would likely lead to the collapse of the fishing sector, as had been the case in many other fisheries around the globe throughout the twentieth century.

Owing to a belief in the maxim that "experts come from out of town," the Peruvian government felt compelled to seek foreign advice by convening international panels of fisheries experts to evaluate the scientific advice that policymakers were receiving from their own scientists. Time after time, panels of foreign fisheries experts issued reports that upheld the research findings and recommendations of their Peruvian colleagues: there were too many boats and too many factories dependent on a dwindling number of fish. In the early 1970s by the time the government finally got the message that it should believe Peruvian scientists, the anchoveta fishery was well on the road to collapse.

1972 and after

1972 was "the year of climate anomalies." Several adverse climate-related events took place that year that had deleterious effects on global food production and global food security. Collectively, these anomalies had a major effect on perceptions about the ability of countries around the world to feed their citizens (Garcia, 1981). The USSR, for example, registered one of its worst shortfalls in grain production, as a result of severe drought. It resorted to major grain imports from the USA, especially wheat and corn, which in turn exacerbated the scarcity of these commodities on the world market. Droughts also occurred that year in Central America, the Sahelian zone of West Africa, India, the People's Republic of China, and in parts of Australia and Kenya. As a result of the 1972 climate anomalies, along with other socioeconomic factors (such as the impacts of, and reactions to, those anomalies), global food production per capita and global food reserves declined for the first time in more than 20 years (Brown and Eckholm, 1974).

In addition to these events and to the decline in Peruvian fish landings, there was a simultaneous decline in fish catches in other parts of the world. Fisheries biologists and national policymakers were forced to rethink their previously held assumption that the oceans would become a major source of food that could supplement food production on land (see, for example, *Mosaic*, 1975; Thompson, 1977).

The 1972–73 El Niño made a bad situation worse. Its impacts on biological productivity, combined with the consequences for the anchoveta population of heavy fishing pressure, contributed in a major way to the collapse of that population and, as a result, to the collapse of the fishing industry in Peru. It also exposed El Niño's impacts on global food production. Because of this El Niño, fishmeal was available only in reduced quantities in the marketplace, and the second preferred choice as a feed supplement of the poultry industry was soymeal. To meet the sharp increase in market demands for feed supplement created by the lack of fishmeal, United States farmers planted soybeans instead of wheat. This was done at a time when a major global food crisis was emerging and, as a result, wheat was in great demand. The price that farmers received for soybeans outpaced what they could get for wheat. The implication of this shift in production from wheat to soybean was serious for the global food situation. At a time when the demand for humanitarian assistance from several developing countries sharply increased, farmers were growing crops as feed supplements for animal consumption instead of producing grain for humans.

Numerous devastating droughts and other weather and climate anomalies worldwide accompanied the 1972–73 El Niño. Rightly or wrongly, many of these were associated by different people with that El

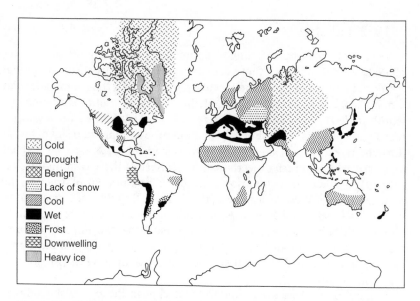

Cold
Drought
Benign
Lack of snow
Cool
Wet
Frost
Downwelling
Heavy ice

Figure 3.9. Global climate anomalies map for the year 1972. This was a year of climate anomalies and of a major El Niño event with substantial impacts on Peruvian fisheries. (From McKay and Allsopp, 1976.)

Niño. These particular impacts prompted some countries to develop an interest in, and respect for, an improved understanding of El Niño events and their consequences, both direct and indirect (Figure 3.9).

The climate anomalies of 1972 sparked a resurgent interest in the study of climate and precipitated the development of a subfield of multidisciplinary research that has since become known as climate-related impacts assessment. Such assessments have focused on the interplay of climate variability and human activities.

Two minor events occurred along the Peruvian coast later in the 1970s. They were followed in the early 1980s by the "El Niño of the century." The 1982–83 El Niño sparked major interest once again in the interactions between oceanic and atmospheric processes in the equatorial Pacific Ocean and the linkages of those processes to climate anomalies around the globe. The magnitude of the 1982–83 event also generated an interest in, and demand for, an improved understanding of El Niño's impacts on human societies and ecosystems among the public and its government representatives. That attention has since been heightened by the two subsequent El Niño events, one in 1986–87 and another that began in 1991–92. Even today, well over a decade after the 1982–83 event, scientists and policymakers seeking to generate interest in El Niño research and impacts, respectively, continue to refer to the 1982–83 event at scientific meetings as "the mother of all El Niño events."

History 2: Interest in the Southern Oscillation

The Peruvian geographer Victor Eguiguren (1895) suggested the existence of linkages between excessively heavy rains and flooding in the northern coastal city of Piura and the warm coastal current called El Niño. Around the same time, the last decades of the nineteenth century, observers of weather variability on the tropical western side of the Pacific Ocean were generating hypotheses about how regional changes in the atmosphere, such as variations in sea level atmospheric pressure, or changes on land, such as changes from one year to the next on the amount of snow cover on the Eurasian land mass, might relate to failure of the monsoonal rains in India. Such changes contributed to recurrent famines in India and to droughts in northern Australia.

In the late 1880s Charles Todd, a government meteorologist in South Australia noted that droughts in India (often referred to as "the failure of the monsoons") seemed to occur at the same time as did droughts in various parts of Australia. He suggested that there were possible connections between the two through the atmosphere. Then, the British empire encircled the globe, and the responsibility for the wellbeing of people in its colonies fell to some extent on Britain. With the major devastation wrought by famine in India in 1877, some researchers focused on trying to identify ways to forecast climate anomalies for the purpose of monitoring food production prospects from year to year.

Also active in the last few decades of the nineteenth century, Norman Lockyer, a British astronomer and meteorologist, sought to identify linkages between solar activity (sunspots) and Indian rainfall. He tried to correlate Indian rainfall with rainfall failures in various parts of Australia, because he too noticed that droughts in northern Australia seemed to occur when there were droughts on the Indian subcontinent.

In 1881 Henry Blanford, the main meteorological reporter for the Indian government, reported to the Indian Famine Commission that he could not find a way to use solar activity to predict weather in India. However, he believed that the increases in snow cover in the Himalayas and on the Eurasian land mass in springtime were responsible for the failure of the monsoonal rains on the Indian subcontinent. Research continues today on the impacts on global climate of Eurasian snow cover.

At the turn of the century, mathematician Gilbert Walker became Director-General of Observatories in India. When he retired in 1924, he returned to England where he turned his attention to attempts to predict the behavior of the Indian monsoon and, later on, changes in sea level

The relationship between El Niño and the Indian monsoon is another topic in which I have been interested. The British government began to pay considerable attention to Indian monsoon variations after the devastating drought of 1877, which we now know was an El Niño year. Another major drought over India occurred in 1918, also an El Niño year, while Walker was the head of the India Meteorological Department. While looking for teleconnections with Indian monsoon rainfall, and not being aware of El Niño, Walker discovered the Southern Oscillation. The relationship between El Niño and monsoons still remains elusive, especially because the major Indian droughts occur six months before the peak phase of El Niño. The quest for predicting monsoons has been further complicated by the fact that the El Niño–monsoon relationship has qualitatively changed during the past 15 years.

J. Shukla, Institute of Global Environment and Society

pressure patterns across the Indian and Pacific Oceans. The particular pattern he observed, which in 1924 he labeled "the Southern Oscillation," was the result of a seesaw-like oscillation of sea level pressure changes at various locations across the Pacific basin. He used pressure records from locations such as Darwin, Australia, Canton Island in the equatorial central Pacific, and Santiago, Chile. Today, researchers use different locations to monitor the same phenomenon – the differences between sea level pressure at Tahiti (French Polynesia) and at Darwin (Australia). The differences between these distant pressure systems (Tahiti minus Darwin) have been converted into an index called the Southern Oscillation Index (SOI). Usually, there is a low pressure system in the region of Indonesia and northern Australia, centered near Darwin. This system brings storminess to the region, providing some parts of Australia, the driest inhabited continent on Earth, with the moisture it sorely needs for agricultural production and for sustaining its settlements, wildlife, and ecosystems. At the same time, there is a high pressure system in the southeastern Pacific, centered near Tahiti. The sea level pressure at each of these two points is related to the other; when it is increasing at one of them, it is usually decreasing at the other. When the sea level pressure at Darwin is low, usually the pressure at Tahiti is high, and vice versa. Therefore, the difference in sea level pressure (Tahiti minus Darwin) is used as an index that characterizes ENSO. Thus, El Niño is related to the negative phase of SOI and the cold event to the positive phase (Figure 3.10).

The SOI has reliably been associated with a number of climate-related events: when the sea level pressure at Darwin increases, the likelihood of drought in the Australian–Indonesian region also increases; when the pressure at Tahiti decreases, the likelihood of more rainfall in the equatorial

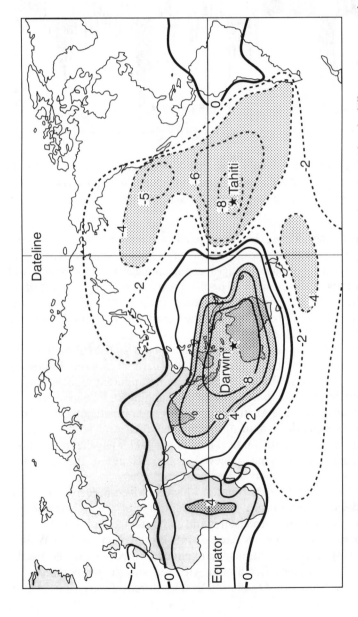

Figure 3.10. Tahiti and Darwin are at opposite ends of the Southern Oscillation's seesaw, and so the difference in pressure between them is used to measure the Southern Oscillation. The numbers represent a statistical measure called the correlation coefficient. The figure shows that the pressure variation at Tahiti is as closely related to the pressure variation at Darwin as are locations near to Darwin, but with the opposite sign (i.e., if the pressure is high at Darwin, it is low at Tahiti and vice versa). (After Rasmusson, 1984.)

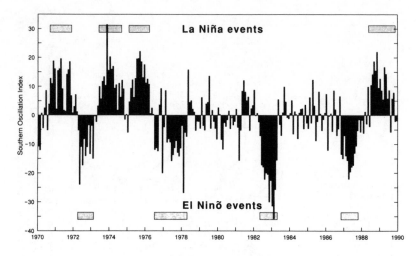

Figure 3.11. Monthly averages of the Southern Oscillation Index and El Niño (warm) events and cold (La Niña) events, 1970–90. El Niño and La Niña years are identified with horizontal bars. (From Nicholls, 1993b.)

central Pacific region increases. These particular sea level pressure changes appear to set the stage for a possible onset of El Niño as suggested in Figure 3.11. Along with pressure changes come changes in wind speed and direction, shifts in the location of pools of warm and cold oceanic surface water, changes in the strength of coastal upwellings, and shifts in the location of biological productivity in the ocean which, in turn, alters the location of various fish populations.

More than 80 years ago, Gilbert Walker used statistical methods to identify weather anomalies linked to the Southern Oscillation. Following on the heels of others, Walker identified many apparent associations or relationships (called correlations) between changing atmospheric pressure patterns at sea level in the Australasian region and rainfall patterns in such distant locations as the Indian subcontinent, Africa, and South America. Because of his strong mathematical background, he was able to use those methods appropriately, and achieved a high standard that was not necessarily followed by many subsequent researchers.

For the most part, these correlations have survived numerous reevaluations and challenges by succeeding generations of researchers using a variety of modern statistical measures and computer technologies. Decades later, to honor Sir Gilbert Walker, Jacob Bjerknes named the Walker Circulation, an important atmospheric circulation pattern in the Pacific that links the Southern Oscillation with sea surface temperatures. In Australia, the Southern Oscillation is monitored closely, not just to

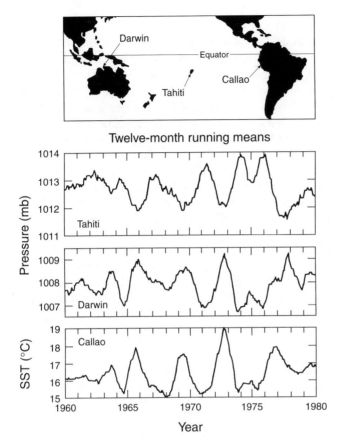

Figure 3.12. Smoothed curves showing changes in atmospheric pressure at Tahiti and Darwin and changes in sea surface temperatures (SSTs) in the eastern equatorial Pacific Ocean at La Punta/Callao, Peru. Callao is the site of IMARPE, the Peruvian Marine Institute. (1 mb = 10^2 Pa.)

anticipate the possibility of El Niño events, but as a tool for forecasting the likelihood of rainfall in various parts of Australia several months in advance. One researcher (see Bacastow *et al.*, 1980) noted that "fully developed El Niño events such as occurred in 1965, 1969, 1972, and 1976 are observed to coincide with a minimum of SOI" (Figure 3.12). He also suggested that there was support for the hypothesis that "all minima of the SOI are accompanied by El Niño-type conditions, even if a fully developed El Niño does not occur." The major point is that, although there is not a perfect one-to-one correlation between observed El Niño events and the SOI, this relationship is *very strong.*

Teleconnections

The word "teleconnection" did not appear in the scientific literature until the mid-1930s (Ångström, 1935). However, the notion behind it, that weather changes at one location might be related to weather changes at other remote locations, had existed for a long time. People have been fascinated since at least the latter part of the nineteenth century by the prospect of identifying linkages among weather events in various parts of the globe. At first, qualitative assessments and scientific hunches were relied upon, as opposed to objective quantitative methods.

Since the turn of the century, the identification of teleconnections has become a subfield of scientific research as well as a pastime for non-weather specialists in their attempts to forecast weather anomalies some months or seasons in advance. Sir Gilbert Walker is perhaps the best known of those who sought to identify such linkages among temperature, rainfall, and pressure anomalies at some distance from each other.

Walker's methods and findings were challenged, if not ignored, by many of his contemporaries right up to the time he died. The way that many in meteorology looked at Walker's teleconnections work (Brown and Katz, 1991) was succinctly captured in an excerpt from his obituary that appeared in the *Quarterly Journal of the Royal Meteorological Society* (**85**, 186) in 1959:

> Walker's hope was presumably not only to unearth relations useful for forecasting, but to discover sufficient and sufficiently important relations to provide a productive starting point for a theory of world weather. It hardly seems to be working out like that.

About a decade later, Walker's contributions to teleconnections research began to overrule the critics. Today, the desire to identify teleconnections with a high degree of reliability is a key driving force behind increased interest in an improved understanding of El Niño processes. Such an improvement could greatly enhance attempts at producing seasonal, interannual, and interdecadal, climate-related forecasts. As one example, Colorado State University professor William Gray uses information on sea surface temperature changes in the equatorial Pacific to construct hurricane-season forecasts for the Atlantic Ocean. He also relies on the "climatology" of El Niño for his projections: his statistical assessments of tropical cyclones in the Atlantic suggest strongly that, during El Niño years, fewer storms can be expected. Following an El Niño, however, there is usually an increase in their number. Yet the search for teleconnections between weather events, as well as between such events and human well-being, is not without its detractors. And even the reliability of some identified El Niño-related teleconnections, such as droughts in Northeast

Brazil or mild winters in the northeastern United States, have been challenged. Clearly, the potential payoff to societies of research that identifies robust teleconnections far outweighs the costs associated with searching for them.

Concluding comments

Ever since Bjerknes identified the physical mechanisms that linked El Niño and Southern Oscillation phenomena, their histories have become intertwined. The most recent situation (i.e., 1991–95) in the Pacific basin has taken the proverbial wind out of the sails of those who overemphasize the tightness of the one-to-one linkage between El Niño in its traditional sense as a local sea surface warming off the Peruvian coast and the Southern Oscillation. Although each phenomenon commands a strong preponderance of regional interest over the other (El Niño captivates the attention of Peruvians, whereas the Southern Oscillation is of major interest to Australians), the Pacific basin-wide El Niño has captured the interest and attention of most El Niño-related researchers and policymakers around the globe.

Section II
The life and times of El Niño episodes

4 The biography of El Niño

The atmospheric circulation in the west-to-east direction in the equatorial Pacific region can be viewed, by analogy, as a stove in a closed room (Figure 4.1). Tribbia (1995, pp. 59–60) described the process as follows:

> The stove in the corner of a room is heating up a portion of air in that room; the warmed air is lighter and rises to the ceiling of the room, crossing laterally toward the window where it cools. It then sinks, reaches the floor, and is sucked back toward the stove.
>
> The tropical atmospheric circulation operates in a similar fashion, if the stove is in the west end of the Pacific basin. To simulate El Niño, however, the stove would have to be moved out to the middle of the room. In this situation one would have rising motion in the middle of the room and sinking motion at both ends. Pressure would be lower in those parts of the room where warm air is rising. In the El Niño-like room, there is low pressure in the center and higher pressure at the sides. Winds flow from high pressure to low.

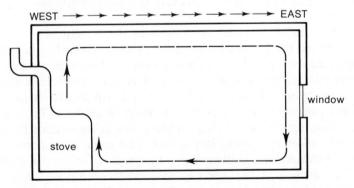

Figure 4.1. Simplified picture of global atmospheric circulation as circulation of air in a closed room. "The heating of the atmosphere drives a cloud system (called a convective cell). Low-level air flows across at the base (from right to left), is heated within the cell and rises and flows out (from left to right) at high levels" (Gill and Rasmusson, 1983). (From Ghil and Childress, 1987.)

The Walker Circulation

Scientists consider "normal" conditions to exist when the following air–sea interactions prevail in the equatorial Pacific. In the western part of the Pacific basin near the equator, there is a very warm pool of water at the ocean's surface, covering an area about the size of the continental USA. That pool extends downward from the surface to a depth of a couple of hundred meters to the zone in the ocean where there is a sharp temperature contrast between the warm waters above and the cold waters below. This zone of sharp temperature change is called the thermocline. The sea level in the western Pacific is higher by a few tens of centimeters than it is at the eastern edge of the basin. This is because of the strong trade winds that flow to the west at the ocean's surface. The trade winds tend to "move" (some say push) water toward the western edge of the Pacific basin.

The large pool of warm water in the western equatorial part of the Pacific Ocean is a major source of heat that warms the atmosphere above it. This warming causes air to rise (by generating convection) which, in turn, produces rain-bearing clouds. As the warmed air rises to even higher levels of the atmosphere, pressure differences between the west Pacific and the east Pacific move the now cooler air to higher altitudes (e.g., the tropopause) and push it toward the eastern part of the Pacific basin. The cool, dry air ultimately descends in the eastern equatorial Pacific. The rule of thumb is that the descending motion of the atmosphere (which is called subsidence) tends to suppress the conditions in the region that could bring about cloud formation and, therefore, rainfall, as depicted in Figure 4.2. The dry air then moves westward near the Earth's surface as a result of wind action. It is warmed by the ocean's surface, from which it is then able to pick up moisture.

Meanwhile, under "normal" conditions in the eastern part of the equatorial Pacific, the thermocline is relatively close to the surface, actually surfacing in the Southern Hemisphere in summer and autumn, whereas in the western Pacific it is at a depth of about 200 meters. The sea level in the east is also several tens of centimeters lower than that in the western part of the basin. Upwelling is strong along the coasts of Ecuador, Peru, and northern Chile, making sea surface temperatures in the eastern Pacific considerably colder than those in the west. This tends to reinforce the atmospheric mechanisms, inhibiting cloud formation and rainfall in the region. The strength of the winds in the equatorial region is heavily influenced by the seesaw-like sea level pressure differences across the equatorial Pacific region.

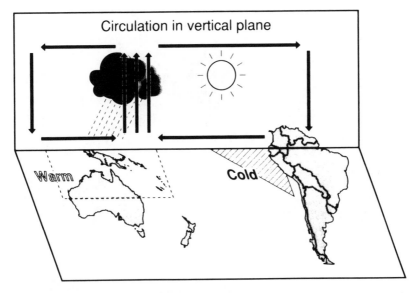

Figure 4.2. A schematic of the Walker Circulation; a cross-section at the equator of the atmospheric and oceanic features in the "normal" (non-El Niño) phase of the Southern Oscillation. (From Nicholls, 1993b).

El Niño and the Walker Circulation

During an El Niño event, the Walker Circulation becomes modified in a major way. The westward-flowing surface winds across the equatorial Pacific basin weaken and in the west they reverse. This enables water in the warm pool in the west to spread eastward. As the warm water shifts eastward, the sea level in the west begins to drop, while sea level in the east increases. With the slowing of the westward winds, the surface waters of the central and eastern Pacific become warmer. As this occurs, the thermocline also begins to shift, moving upward toward the ocean's surface in the west and deepening in the central and eastern equatorial Pacific. As the thermocline moves away from the surface along the Peruvian coast, upwelling continues but the water brought up to the surface is warmer and less rich in nutrients.

Meanwhile, the water in the western equatorial Pacific becomes a few degrees cooler, as water in the central and eastern Pacific warms up. Because convective activity follows the warm waters at the sea's surface, clouds increase in the central and eastern Pacific, while they decline in the west. This displacement in convective activity leads to droughts in Australia and Indonesia, typhoons in the central Pacific, and heavy rains along the

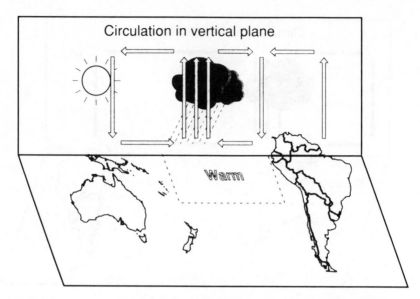

Figure 4.3. A schematic of the Walker Circulation; a cross-section of the atmospheric and oceanic features in the El Niño phase of the Southern Oscillation. (From Nicholls, 1993b.)

normally arid coast of Peru (Figure 4.3). These conditions can last from 12 to 18 months, until the surface winds once again begin to strengthen and flow westward, causing warm water to flow back toward the region of the western Pacific warm pool. The sea levels at both ends of the basin begin to change direction as does the depth of the very important, but out-of-sight, thermocline. Strong upwelling returns to the equator and to the eastern Pacific boundary of coastal Peru.

Figures 4.2 and 4.3, respectively, depict the west-to-east interaction between the atmosphere and the ocean in the equatorial Pacific under "normal" conditions, and under El Niño conditions. The precise timing of any particular El Niño event may not be well known, although there are several hypotheses about how to detect it. Once started, however, the processes that keep it going, as well as the processes that end it, are better known. In fact, once an El Niño has started, scientists usually have a good idea of its sequence, including its demise.

The El Niño process over time

The phases of El Niño

Early research on El Niño seemed to suggest that there was only one kind of El Niño. It would start along the east coast of South America when

Jacob Bjerknes was, of course, the scientist who recognized that the interplay of ocean and atmosphere was responsible for the generation of El Niño. My idea that El Niño is a heat relaxation of the equatorial Pacific ocean expands on his work and emphasizes the importance of ocean circulation in this process. The west Pacific warm pool slowly grows in size and depth, because ocean circulation is not capable of removing all the accumulated heat from this area. As the warm pool increases, atmospheric convection starts to move east toward the central Pacific and the associated winds trigger the equatorial Kelvin wave in the ocean. This wave moves heat eastward [along the equator] and poleward [along the eastern boundary] and thus removes heat from the area of the warm pool. This heat relaxation process determines the length of an El Niño cycle.

Klaus Wyrtki, University of Hawaii

sea surface temperatures increased, and would propagate westward toward the central Pacific. This picture of El Niño, however, may have been false because it was made up by averaging several El Niño events together. Averaging takes away any indication of how and in which way each event was unique, when compared to each of the others.

Figure 4.4 was produced by American meteorologists Eugene Rasmusson and Thomas Carpenter to define what was called the canonical (i.e., typical) El Niño. The 1982–83 event has been added to their original figure. The 1982–83 event differed in several respects from the earlier ones depicted here, prompting them to consider more closely the differences between El Niño events.

Now that scientists are looking more closely at each specific event, they can identify those features that are similar, as well as those that are unique to any given event. It is not difficult to see, for example, that El Niño events can form in at least two different ways, based on whether the sea surface temperatures first heat up either in the eastern or the central equatorial Pacific. Such information can help to identify the various forms that an El Niño event might take.

As Figure 4.4 shows, El Niño events have a life cycle. Rasmusson and Carpenter identified several stages of El Niño's development: antecedent, onset, peak, transition, and maturity. The components of this life cycle are depicted in Figure 4.5, which was based on the set of El Niño events that took place before 1982–83, the same set of events they used to define a canonical El Niño.

With the realization after 1982–83 that El Niño events can begin at different times of the year, tying the phases of the life cycle to specific months of the year became more troublesome. Sometimes the warming

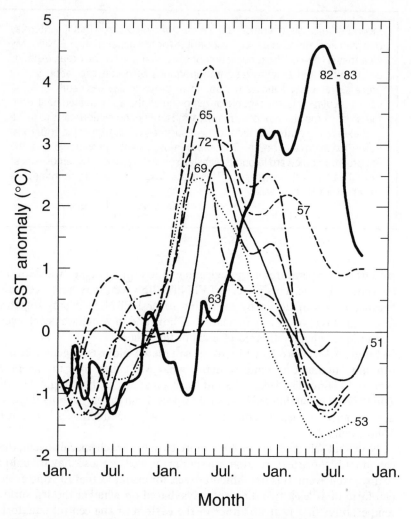

Figure 4.4. A composite El Niño sea surface temperature (SST) anomaly pattern. The sequencing of the 1982–83 El Niño event does not match the composite of previous El Niño events. (From Rasmusson and Carpenter, 1982.)

along the Peruvian coast occurred before the warming in the central Pacific, and sometimes it followed it.

Even though there are different types of El Niño event based on the timing of onset and location of anomalous increases in sea surface temperatures, they still evolve in a similar way, undergoing the same life cycle sequence of growth to decay. Australian meteorologist Neville Nicholls characterized the El Niño process using slightly different terms: a

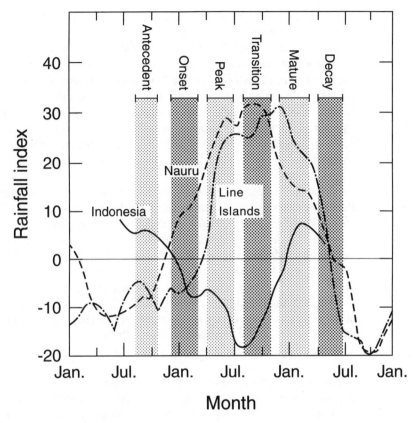

Figure 4.5. *Development of rainfall over time at three locations in the equatorial Pacific Ocean: near Nauru (167° E), and the Line Islands (near 160° W) for the El Niño events before 1980. Note the out-of-phase relationship between Indonesian and Nauru rainfall anomalies, and the eastward movement of the rain anomaly from Nauru to the Line Islands. The rainfall tracks the anomalously warm sea surface as it moves eastward into the Pacific. (From Tomczak and Godfrey, 1994.)*

precursor phase, an onset phase, a phase when anomalous conditions grow and mature, and a phase during which those anomalous conditions decay (Nicholls, 1987).

(a) *Precursor phase* One could argue that the precursor phase begins at the end of a cold phase event, when sea surface temperatures have returned to near normal, usually following, and followed by, an extreme cold event.

Let us assume that the precursor phase begins just after the height of the cold phase. For reasons that are not completely understood, the strong westward-blowing winds (the trade winds) begin to weaken, the sea level in the western Pacific reaches its peak, and the sea level along the western coast of South America reaches its minimum for that cycle. With a weakening of the surface westward-flowing winds, equatorial and coastal upwelling begins to reduce, and sea surface temperatures in the central and eastern equatorial Pacific begin to warm up. This is the transition phase, moving out of a cold event and toward a warm event of unknown magnitude.

(b)   *Onset phase*   Around December of each year (the beginning of Peruvian summer), there is a seasonal slackening of the winds off the coast of Peru and Ecuador. At that time, the upwelling of cold water along the coast slows down, and surface water heats up, lasting until March or so. If that seasonal warming were to continue into April and May, it is likely that the onset of a warm event of some magnitude would be under way. However, the onsets of some El Niño events (for example, 1982–83) have occurred later in the year (after August), when the sea surface temperatures had already returned to normal in April following the seasonal warming.

(c)   *Growth and maturity phase*   As the months proceed, the sea surface temperatures in the central and eastern equatorial Pacific become increasingly warmer, and upwelling ceases to bring nutrient-rich cold water into the sunlit zone of the ocean surface. The sea level pressure in the South Pacific (near Tahiti) drops, and the pressure at Darwin increases. With the weakening of the westward-flowing winds and the strengthening of the eastward-flowing winds, the area covered by warmed surface water expands in the central and eastern Pacific. Sea surface temperatures can increase from 1 °C to 4 °C or more (as happened in 1982–83). The sea level in the western Pacific drops a few tens of centimeters, while sea level in the eastern equatorial Pacific increases.

(d)   *Decay phase*   This phase begins once maximum sea surface temperatures have been reached in the central and eastern equatorial Pacific, and surface temperatures begin to respond to changes in wind speed and direction across the basin. The thermocline begins to move in the opposite direction (once again becoming deeper in the west and shallower in the east), and the warm water pool begins to thicken in the western part of the basin.

Westward-flowing winds again begin to strengthen and eastward-flowing winds weaken. Coastal and equatorial upwelling begins to strengthen, bringing more cold, deep water to the ocean's surface. And the cycle (i.e., oscillation) begins once again toward the onset of a cold phase.

Scientists often refer to the year of the actual onset of El Niño as year 0, the year before it as year −1 and the year following the onset as year +1. A few key environmental changes bear monitoring: sea surface temperatures, winds, sea level, sea level pressure, and the thermocline. However, processes in the Pacific Ocean and atmosphere and the results of their interactions are quite complicated. Some scientists may favor watching changes in sea surface temperatures in a specific area as *the* leading indicator of an El Niño event, while others might consider monitoring the surface winds to be more important, and still others focus directly on the Southern Oscillation.

There is not just one type of El Niño. The El Niño that scientists had considered to be typical (i.e., the ones that occurred in the post-war years until 1982) differed in several respects from those after 1982. Thus, despite progress made in understanding the El Niño phenomenon in general, it has been difficult for scientists to identify the precise timing of the onset or the decay of a warm event or of a cold event, or the geographic scope or intensity of either. Researchers will eventually identify various types of El Niño, thereby generating more confidence of potential users of El Niño information in El Niño forecasts.

Niño regions

Scientists have identified four regions in the equatorial Pacific that they consider to be worthy of special attention with regard to the observations and monitoring of El Niño processes. These are referred to as Niño1 to Niño4. Their approximate locations are shown in Figure 4.6.

Each region provides different kinds of information about either El Niño or the Southern Oscillation.

- Niño1 is the region of coastal upwelling off the coasts of Peru and Ecuador. It is sensitive to changes in the ocean and the atmosphere, both seasonally and especially during El Niño episodes. Coastal upwelling processes in Niño1 are particularly sensitive to changes in air–sea interaction in the central and eastern equatorial Pacific.
- Niño2 represents the Galapagos Islands region of the equatorial Pacific. Equatorial upwelling processes in this area are also sensitive to seasonal, as well as El Niño-induced, changes in the marine environment. Niño2 is a transition zone between the

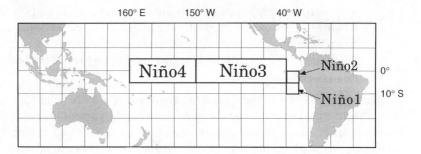

Figure 4.6. Map depicting four regions (referred to as Niño1, Niño2, etc.) in the equatorial Pacific Ocean identified as important locations for monitoring the wind and sea surface temperature changes associated with the El Niño process.

central and eastern equatorial Pacific, sensitive to changes in either Niño1 and Niño3, or both.

• Niño3 is in the central equatorial Pacific, where there is a large El Niño signal but not much sensitivity to seasonal changes in air–sea interaction. It is in this region where information on changes in surface winds has been used by Mark Cane and Stephen Zebiak to project the likely onset of El Niño events. According to Cane (1991, pp. 357–8), "a warming in this region is thought to influence the global atmosphere strongly. It is probably the best single indicator of an ENSO episode likely to affect global climate."

• Niño4 encompasses part of the western equatorial Pacific known as the warm pool. Here, sea surface temperatures are the highest in the Pacific. During an El Niño event, there is a relatively small change in sea surface temperatures. However, that small change is important, because the warmest water and the cloud-producing processes that tend to follow it move toward the central Pacific.

Waves in the ocean

Hidden from the naked eye, waves exist inside the ocean, several meters to hundreds of meters below the ocean surface (i.e., an internal wave). No matter how hard one might try to avoid the mention of these waves, a reliable understanding of El Niño cannot be gained without it. Scientists have identified two types of internal wave: the Kelvin wave and the Rossby wave.

Kelvin waves are created by winds blowing over the ocean surface from the west along the equator. Before a warm event develops, eastward-flowing winds increase over the area of warm sea surface temperature to the east of New Guinea (the region referred to by scientists as the warm pool).

El Niño and the Galapagos Islands

The Galapagos Islands, off the coast of Ecuador, are in an oceanic transition zone between the central and eastern equatorial Pacific. The islands are directly affected by both weak and strong El Niño events, whether the warm water initially forms in the central Pacific toward the east coast of Ecuador and Peru, or whether it initially forms along the coastal area toward the central Pacific.

The Galapagos Islands are considered to be quite exotic with regard to flora and fauna; thus, the area has gained notoriety as a tourist attraction. The islands are normally dry for half the year (June to December) and relatively wet for half the year (January to May). This enables seabird populations, marine and land iguanas, penguins (yes, penguins exist at the equator), and giant tortoises to flourish. However, when an El Niño event develops, flora and fauna on the islands change. Some researchers have noted that the Galapagos "are perfectly located for detecting disburbances propagating in the equatorial waveguide" (this is the region in the Pacific Ocean that is delimited by the 5° north and south latitudes) (Kogelschatz et al., 1985, p. 96).

As sea surface temperatures increase during an El Niño event, so does precipitation. During the 1982–83 event, heavy rains fell over the Galapagos from November 1982 to July 1983 (Kogelshatz et al., 1985, p. 98). For example, rainfall on one of its islands, Santa Cruz Island, reached 3224 mm for the year, whereas the five-year average (1965–70) was 200 mm per year.

According to oceanographic researchers, one of the most significant changes that occurs in the Galapagos during an El Niño is the depletion of nutrients in the water. In reflecting on the 1982–83 event, some have noted that "this depletion of new nutrients necessarily reduced the new primary production. . . . After a period of time the decrease in new primary production must have caused proportional reductions in the growth and reproductive success of zooplankton, fish, birds, and marine mammals" (Kogleshatz et al., pp. 119–21). In his recent book on the natural history of the Galapagos Islands, Michael Jackson spoke of the adverse effects during the 1982–83 El Niño:

> Life on land burgeons but seabirds, which depend on the productive cooler waters, may experience dramatic breeding failures. . . . Seabird colonies, such as the blue-footed booby colony on Española [one of the dozen or so islands that make up the Galapagos archipelago], temporarily vanished, sea lions and the seals died of hunger, vegetation went rampant, and Darwin's finches were able to breed at an unusually high rate. (Jackson, 1993, p. 312)

M. H. Glantz

These so-called "westerly wind bursts" produce two effects: they move warm water eastward from the warm pool to the central equatorial Pacific, generating the warm sea surface temperatures observed there, and they also produce Kelvin waves, causing a lowering of the thermocline and an increase in sea surface temperature over the eastern Pacific. Changes in the thermocline cause small changes in sea level (which can be measured by satellite). Although the thermocline cannot be measured directly by satellite, satellites can measure, as proxy information, changes in sea level that have been correlated with changes in the thermocline. Kelvin waves can change the depth of the thermocline by 30 meters or more, and the sea level by tens of centimeters. More specifically, this changes the volume of warm water in the western Pacific's warm pool, as the thermocline becomes shallower than normal, while in the eastern part of the basin, the volume of warm water increases. Kelvin waves are the mechanism that is said to be responsible for the popular notion that warm water sloshes back and forth (from west to east and back again) across the equatorial Pacific basin, as water does in a bathtub each time the person in the bath moves.

The net effect of a series of Kelvin waves is to raise the sea surface temperature systematically over much of the equatorial Pacific. Thunderstorm activity, following the warm water, and usually located over Indonesia and the warm pool, also moves eastward into the central and eastern Pacific. This weakens the trade winds even more as the Walker Circulation moves eastward, causing further warming of the sea surface temperature, and so on. Scientists call this interaction a "positive feedback" mechanism between the atmosphere and ocean, one that results finally in El Niño conditions.

A Kelvin wave takes about $2\frac{1}{2}$ months to travel across the Pacific basin, a distance of one-third of the circumference of the Earth. Once forced, Kelvin waves move eastward, independently of the season. Although it has been centrally implicated in starting off an El Niño, a Kelvin wave does not necessarily lead to an El Niño event.

When the Kelvin wave "hits" the coast of South America, it generates coastal Kelvin waves that propagate both to the north and to the south along the western coasts of North and South America. It also generates what is known as a *Rossby wave*, a westward-moving internal wave that travels at one-third the speed of a Kelvin wave. It takes about nine months for a Rossby wave to cross the Pacific. Rossby waves depress the thermocline in the western Pacific region. Some people think a Rossby wave begins the process of decay of an El Niño, suggesting that the onset of an El Niño carries with it the seeds of its own destruction. While such internal waves are not visible to the naked eye, there are ways to identify their existence using indirect measures, as suggested in Figure 4.7.

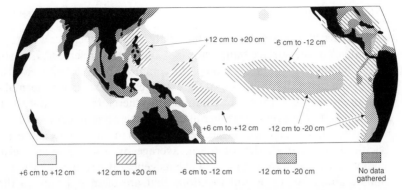

Figure 4.7. Sea level change between 1987 and 1988 (1988 data minus 1987 data). Scientists use maps like this to show sea level changes in the Pacific between an El Niño year (a warm event) and a La Niña year (a cold event). It demonstrates the recovery of the sea level from the 1986–87 El Niño into the opposite phase (sea level drops in the east from being high in 1987 (El Niño) to being low in 1988); in the west it goes from being low in 1987 to being high in 1988. This is like the seesaw pattern one sees with the Southern Oscillation. This map also shows that the drop in sea level in the east is trapped near the equator and along the coast (where it moves poleward in each hemisphere). The equatorial and coastal signal is characteristic of equatorial Kelvin (internal) waves. It is also interesting to note the connections to higher latitude implied by positive sea level changes in the western part of the Pacific. (After Koblinsky et al., 1992.)

A lingering El Niño?

Very recently, some oceanographers claim to have discovered that the impacts of El Niño events on the oceanic environment do not dissipate within a few years, as had been believed. They have suggested that the consequences of the 1982–83 El Niño were still being felt in the north-western Pacific Ocean as recently as 1994 in the form of a Rossby wave. This very slow-moving wave, according to researchers Gregg Jacobs and his colleagues (1994), has affected the location of the northern extension of the Kuroshio current, a warm current in the Pacific that functions similarly to the Gulf Stream in the North Atlantic. The Kuroshio current determines, to a large extent, the climate of the countries bordering the North Pacific. This "lingering" Rossby wave has been suggested to have pushed the warm Kuroshio current further to the north, affecting ocean temperatures and perhaps weather patterns on the North American continent as well. While this view has been received with interest by some researchers, it remains highly speculative, highly controversial, and difficult to imagine. The Rossby wave is there; its alleged influence, however, is unclear.

Cold events

Warm events (El Niño) are only part of the cycle taking place with regard to changes in sea surface temperatures. Cold events (referred to by some researchers as La Niña (see box on p. 16)) complete the cycle, with warm and cold events generally appearing at the extremes of the seesaw pattern of sea level pressure, the Southern Oscillation. It is, therefore, important to view El Niño and cold events together as part of the same phenomenon.

During cold events, sea surface temperatures in the eastern and central Pacific decrease from the long-term average by a few degrees Celsius. Today, researchers argue that cold events generally produce weather and climate anomalies in distant locations that are often opposite to those produced in the same region by El Niño. For example, cold events are believed to be associated with good rains and favorable agricultural production in Indonesia, Australia, and Northeast Brazil, whereas El Niño has been associated with droughts in these regions.

Researchers for the most part have shown less interest in cold events than in El Niño. Perhaps this is because such events are associated with periods of weather and climate conditions that are perceived as normal in various regions. Only one cold event occurred between 1975 and 1988. Their paucity in that period has most likely reinforced the lack of interest in the potential consequences of cold events.

From the perspective of societal impacts, it also appears that cold events, or, more precisely, anything that is not an El Niño situation, have generally been viewed as "normal." This view, at least to the public, is reinforced by such media statements as "El Niño is ending and the weather will return to normal."

5 The 1982–83 El Niño: A case of an anomalous anomaly

The *American Heritage* dictionary defines an anomaly as a deviation from the normal order and as something unusual or irregular. El Niño is an anomaly in terms of changes in sea surface temperatures and in sea level pressure. The 1982–83 El Niño can, therefore, be described as an anomalous anomaly. In addition to being unusual, it is considered by the scientific community to be the most extreme this century.

The sea surface temperatures associated with this El Niño were way above normal in the central and eastern equatorial Pacific (reaching 4 deg.C or more above normal in some areas). Those concerned about natural disasters and their socioeconomic impacts also considered it to have been a very extreme event, because of numerous destructive climate anomalies that occurred around the world at the time. In the wake of this particular El Niño event, many world leaders, the public, and the media, for the first time, were forced to pay more consistent attention to this phenomenon.

In 1982, several months prior to the onset of the 1982–83 El Niño, some researchers published an article on what they considered *the* "typical" El Niño. The canonical El Niño, as they called it, was in fact a composite of the features of several El Niño events that took place from the early 1950s to the late 1970s. Until then, it appeared that El Niño events had generally followed a similar pattern of growth and development from onset to decay. In less than a year, however, that notion was challenged with the occurrence of an unexpected (by most, but not all, researchers) out-of-phase onset of the 1982–83 event. According to Ed Harrison and Mark Cane,

> The warm event in the Pacific in 1982–83 was unusual in many respects. Rather than exhibiting surface warming first along the northeast coast of South America in the spring, sea surface temperatures first significantly exceeded climatological values along the equator in the eastern central Pacific during late summer.
>
> (Harrison and Cane, 1984, p. 21)

In addition, Rasmusson and Arkin (1985, p. 183), in reviewing what had happened in 1982–83, wrote that the timing of the warming of water in the central Pacific was typical and that the warming along the Peruvian coast followed instead of preceded that warming. In any event, the coastal warming occurred at the normal time of the year.

The 1982–83 El Niño differed in both timing and location from the set of post-World War II El Niño events used to compile the characteristics of a "typical" event. For example, anomalously warm sea surface temperatures appeared first in the central Pacific instead of off the coast of Peru. The warm sea surface temperatures moved eastward toward the South American coast, instead of first appearing along the coast and then moving in a westward direction away from it. It emerged later in the year (between June and August) than the expected typical El Niño. The winds along the Peruvian coast did not weaken as expected when the El Niño began, even though a weakening of the westward-flowing winds was considered to be a necessary, although not sufficient, condition to spark El Niño's onset.

Another unique aspect of the 1982–83 event was the breadth and severity of its ecological and societal impacts. In addition, because it was "late," it was unexpected by most scientists. Societies that might have been affected by it were caught off guard, even in those societies where El Niño impacts have a strong likelihood of occurrence because of El Niño's teleconnections. The 1982–83 El Niño demonstrated to the scientific community, and to the federal agencies that funded El Niño research at that time, just how much about this potentially devastating natural phenomenon was yet to be learned.

By all published accounts, most scientists involved in El Niño research had failed to recognize the potential for the development of the 1982–83 El Niño for several months. Typifying the situation, a senior El Niño researcher had been in Peru in August 1982 and, as he left, he announced that there would be no El Niño that year as the various early warning indicators that he relied on to forecast an event were apparently not evident. Within the next few months at meetings of El Niño researchers in Miami, Florida, and Princeton, New Jersey, scientists in attendance also concluded that there would be no El Niño event that year. Within a matter of weeks, they were proven wrong in their assessment.

There were, however, two notable exceptions. Australian meteorologists noted that the Southern Oscillation Index was extremely low in mid-1982, a strong signal of an impending El Niño. Eugene Rasmusson from the USA suggested in June that an El Niño was likely to emerge later that year. The scientific community officially noted that it had failed to recognize the onset of the 1982–83 El Niño. A scientific report summarized what had happened:

The planning of TOGA [*Tropical Ocean-Global Atmosphere*] was just getting under way in November 1982, when the strongest ENSO event thus far this century caught the scientific community by surprise. Unlike its predecessors during the previous three decades, the 1982–83 warm episode was not preceded by a prolonged "buildup phase" with strong trade winds along the equator, and it did not exhibit what had come to be viewed as the typical "onset phase" around April, characterized by El Niño conditions along the South American coast, which would later spread westward across the basin. In retrospect, it is apparent that the first indications of a major warming should have been evident in July and August 1982, when anomalous equatorial westerlies were observed to develop in the central Pacific, accompanied by strong sea level pressure rises at the western end of the Pacific basin. El Niño conditions did not become apparent along the South American coast until November, by which time the basin-wide warming had nearly reached its peak.

(NRC, 1990, pp. 11–12, and summarized from Rasmusson and Wallace, 1983)

In response to the 1982–83 event, the scientific community declared that, like snowflakes, no two El Niño events were alike. The truth, of course, lies somewhere in between these opposing views. Most likely, there are types of El Niño events based on sets of characteristics, including, for example, where and when the sea surface temperatures begin to increase.

Impacts associated with the 1982–83 El Niño

Most of the major weather anomalies occurring in 1982 and 1983 around the world, especially droughts and floods in the tropics, were linked by one observer or another to the occurrence of an El Niño. Several articles, maps, and charts relating to El Niño appeared in the popular press, suggesting the extent of the worldwide, continent-wide, national, and local impacts of this El Niño. Caution must be used, however, in attributing any particular anomaly or impact to a specific El Niño. Furthermore, the severity of societal impacts will vary according to the level of societal vulnerability to such extremes. Climate-related anomalies can also result from a variety of local and regional conditions, even in the absence of El Niño events. The following examples of the alleged societal impacts of the 1982–83 El Niño are taken from newspaper reports:

- Indonesia was plagued with severe drought, resulting in reduced agricultural output (especially rice), famine, malnutrition, disease, and hundreds of deaths. This drought came at a bad time, in the sense that this country had been making great strides toward self-sufficiency in food production. In the few years immediately preceding the 1982–83 El Niño, Indonesia had begun to emerge as a rice exporter. This drought, however, coupled with worldwide recession, huge foreign debts, and declining oil revenues, set back Indonesia's economic development goals for the near term.

- In 1982–83, Australia was in the midst of its worst drought this century up to that time. Agricultural and livestock losses, along with widespread bush fires mainly in the southeastern part of the country, resulted in billions of dollars of lost revenue. The El Niño exacerbated this situation. An Australian journalist wrote that the drought was not just a rural catastrophe, it was a national disaster.

- The eastern part of the USA was favorably affected by its warmest winter in 25 years and the fewest hurricanes of the century up to this date. According to an estimate by the National Oceanic and Atmospheric Administration, energy savings were on the order of $US500 million. (The opposite was the case, however, during the cold winter that accompanied the 1976–77 El Niño.) Also in 1982–83, the USA was adversely affected by devastating coastal storms and mudslides along the southern California coast, flooding in the states bordering the Gulf of Mexico, and drought in the north central states, reducing corn and soybean production. Salmon harvests along the United States Pacific Northwest coast were also down sharply due to reduced coastal upwelling and a general warming of the ocean's water, which pushed salmon populations further north into Canadian waters and into the hands of Canadian fishermen.

- South America experienced many and varied impacts. In addition to the highly publicized damage to infrastructure such as roads, railroads, and bridges, and agricultural production in Peru and Ecuador as a result of heavy flooding during the 1982–83 El Niño, there were severe droughts in southern Peru and Bolivia. A major drought continued in Northeast Brazil, adversely affecting food production, human health, and the environment. The drought prompted migration out of the region into the Amazon and into the already crowded cities along the coast and to the south. There were also destructive floods in southern Brazil, northern Argentina, and Paraguay.

- Large expanses of Africa were affected by drought. For example, the West African Sahel was, once again, plagued by a major drought. Although the human and livestock deaths resulting from this drought appeared to be lower than those that occurred during the 1972–73 El Niño, the situation with food production was considered extremely poor. The view that the Sahel was in the midst of a long-term trend of below-average rainfall that began in 1968 gained some credibility.

- Southern Africa has witnessed some of its worst droughts, including that of 1982–83, during this century. For example, in 1983 the Republic of South Africa, a major grain producer in the region, was forced to import about 1.5 million tonnes of corn from the USA to replace what was lost in their drought. Zimbabwe, a regional supplier of food, was also devastated by drought and was forced to appeal for food assistance from the international community. Likewise, Botswana, Mozambique, Angola, Lesotho, and Zambia, and the so-called Black National Homelands in the Republic of South Africa had their economies devastated by the drought of 1982–83.

In addition to these impacts, the El Niño of 1982–83 was blamed for droughts in Sri Lanka, the Philippines, southern India, Mexico, and even Hawaii, along with severe, unseasonal typhoons in French Polynesia and Hawaii. It was also credited with having a role in suppressing hurricane activity along the Atlantic seaboard. In 1983, many of these events were record-setting extremes: the worst typhoon, the most intense rainfall, the warmest winter, the longest drought, and the fewest hurricanes making landfall on the eastern USA, all occurred in this year.

El Niño has also been associated with indirect societal and environmental effects. However, indirect effects are even more difficult to attribute to an El Niño, as they could be the result of other causes. In 1982–83, these effects took the form of dust storms and bush fires in Australia, the Côte d'Ivoire, and Ghana. In the USA, the 1982–83 event was blamed for such health effects as encephalitis outbreaks in the East (the result of a warm, wet spring providing the proper environment for mosquitos), an increase in rattlesnake bites in Montana (hot, dry conditions at higher elevations caused mice to search for food and water at more densely populated lower elevations; the rattlesnakes followed the mice), a record increase in the number of bubonic plague cases in New Mexico (as a result of a cool, wet spring that created favorable conditions for flea-bearing rodents), an increase in shark attacks off the coast of Oregon (because they followed the unseasonally warm sea temperatures). Even an increase in the incidence of spinal injuries along California's coast was blamed on El Niño (as a result of swimmers and surfers being unaware that the floor of the ocean along the coast had been changed as a result of the violent wave action that accompanied coastal storms).

Figure 5.1, based on information compiled by the United States government's National Oceanic and Atmospheric Administration, depicts the hypothesized socioeconomic impacts and costs associated with the 1982–83 event. It is important to note that there are differences among the various El Niño impacts maps. This particular one identifies impacts and

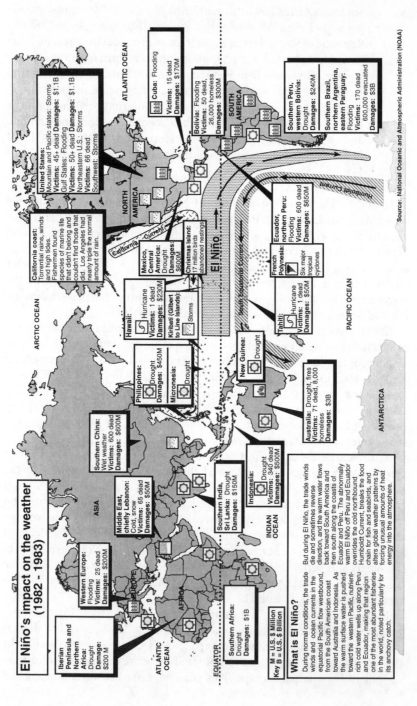

Figure 5.1. National Oceanic and Atmospheric Administration's compilation of the damages only of the 1982–83 El Niño. (From NOAA/OGP.)

costs of worldwide climate anomalies and attributes them to El Niño. Other maps simply identify anomalous climatic conditions that have occurred during an El Niño event. In time the research community hopes to sort out those teleconnections (i.e., distant worldwide climate-related anomalies) that might rightly be attributed to an El Niño event from those with more tenuous linkages.

Positive impacts

There has been an overwhelming tendency to focus on the adverse impacts of El Niño on human activities. However, with regional shifts in temperature and precipitation, one can expect that some regions as well as some human activities will benefit from those shifts.

In 1588 José de Acosta published in Spain an account of his travels, which included a visit to Peru. He reported on the coastal activities of the local populations, describing the balsas, the boats made of balsa tree logs bound together.

> Rowing up and downe with small reedes on either side, they goe a league or two into the sea to fish, carrying with them the cordes and nettes. ... They cast out their nettes, and do there remaine fishing the greatest parte of the day and night, untill they have filled up their measure with which they returne well satisfied. Truly it was delightfull to see them fish at Callao off of Lima, for they were many in number.

This is an interesting account, suggesting a great abundance of fish relative to the technology available to fishermen at that time. To meet their needs from these marine resources, fishermen exploited the fish at levels that were well below the maximum levels that the various fish populations could withstand and still survive. During El Niño events, pockets of cold upwelled water appear in the midst of a broad swath of warmer sea surface water, and those pockets are close to shore. One could argue effectively that, during El Niño, fish in those cold water pockets would become easier to catch and require less effort on the part of the fishermen.

Today, El Niño suggests "bad times" for some species of fish (especially anchoveta) and for the fishermen along the Peruvian and northern Chilean coasts, but, as one can see from Acosta's remarks, Peru's coastal waters were teeming with living marine resources hundreds of years ago. So what is the difference? Is it just El Niño? That could not have been the only pressure on the living marine resources. There have been such events for thousands of years and there have remained an abundance of fish to catch. What has changed is society's ability to capture fish in great numbers, numbers that go much beyond just satisfying the needs of humans for food. Most recently, Princeton University oceanographer George Philander (1995), commented on what he considered to be a misperception of El Niño. He reminded other researchers that local fishermen and populations along the coast of Peru had traditionally referred to El Niño years as "years of abundance." From their perspective, El Niño was a good thing. There are known positive as well as negative impacts of El Niño along the western coast of South America: e.g., the impacts of El Niño on phytoplankton, on the benthic community (those marine resources that live on the bottom of the ocean's floor), on increases in the sardine, jack mackerel, and scallop populations, on seal populations, and so forth. In fact, an American geologist, writing about the coastal deserts of Peru at the end of the 1800s (Sears, 1895), recorded many of the positive impacts of El Niño-related heavy rains in an otherwise perennially hostile arid environment.

Almost a century later, German researcher Wolf Arntz witnessed some examples of those positive changes that accompanied the 1982–83 El Niño:

> In many parts of the coastal desert, rains and unusually strong fogs [*led*] to a hitherto unknown outburst of vegetation that covered wide areas with carpets of flowers for several months, enabling local settlers to raise cattle, sheep, and goats. ... Apparently, the seeds and bulbs of many plants survive in the desert for many decades until a strong El Niño creates appropriate conditions for the type of explosion observed this year.
>
> (Arntz, 1984, p. 37)

Limits of utility of the 1982–83 (or any other) case study

As the biggest event in a century, the 1982–83 El Niño was instrumental in prompting governments around the Pacific basin, especially those of the USA, Canada, Japan, Australia, and Peru, to pay more attention in general to variations in sea surface temperature and sea level pressure in the Pacific, and specifically to El Niño and to the Southern Oscillation.

Today, just about everyone who knows about El Niño seems to refer back to the 1982–83 event in much the same way that recent droughts in North America continue to be compared to those that occurred during the Dust Bowl days in the Great Plains region in the mid-1930s. The reasoning goes as follows: because it was the biggest El Niño, its "signal" and some of its impacts were quite apparent. In science jargon, the El Niño *signal* surpassed the *noise* surrounding it. This is not often the case for weak or even moderate El Niño events. Where members of the public have become aware to some degree of the El Niño phenomenon, they tend to believe that future El Niño events are likely to yield similar teleconnections (weather anomalies), more or less in the same locations where they occurred during 1982–83.

Although some anomalous events, such as drought in Northeast Brazil, flooding along the northern Peruvian coast or in southern Brazil, or drought in parts of Australia and in Indonesia, have relatively strong and reliable linkages with El Niño, many others, such as floods along the California coast, plague outbreaks in the southwestern United States, or fires in Malaysia or Indonesia, are not to be expected during each future El Niño. Thus, while the 1982–83 El Niño is a very informative, instructive and, therefore, useful case study, lessons drawn from it must be used with great care. It is only one case for study, albeit an important one.

One researcher observed that

> El Niño is but one of many influences upon the global atmosphere. During any particular event the other processes may reinforce or obliterate the distant influence of El Niño. Climate anomalies observed in other ocean basins during El Niño can result from anomalous surface wind in those regions. The wind anomalies may or may not be "caused" by the El Niño.
>
> (Hansen, 1990, p. 6)

It is important to reiterate that the actual level of impact of an El Niño also depends on the local and regional climatic settings that exist at the time of its onset. Even if two El Niño events were to have the same physical characteristics, their consequences could vary widely from one place to another, and could even vary in the same location at different times, depending on a host of societal as well as environmental factors at the time they occur.

The 1982–83 event is probably the most studied El Niño, having been analyzed by meteorologists, ecologists, hydrologists, fisheries experts, biologists, ornithologists, oceanographers, social scientists, and economists, among others. Few assessments, however, focused on its societal impacts in the tropics and elsewhere. Several workshops were held in the mid-1980s in Ecuador, Peru, Chile, and elsewhere to identify the regional impacts of the 1982–83 El Niño. Nevertheless, a major benefit to the *scientific* community of the devastating 1982–83 event was the recent development of a major research effort called TOGA (Tropical Ocean–Global Atmosphere) to better understand various air–sea interactions across the tropical Pacific Ocean.

Many members of the scientific community apparently believe that the devastation associated with the 1982–83 event provides overwhelming evidence in support of the need for El Niño research. However, I believe that the 1982–83 El Niño is becoming overused as an example of typical El Niño-related devastation. In the last decade, two or more El Niño events have taken place: in 1986–87 and one or more in the 1991–95 period. Yet, no major objective accounting of the societal and economic impacts and consequences of those events has been undertaken.

Furthermore, in light of our improved scientific understanding of the El Niño phenomenon, one might realistically assume that the cost (in terms of constant dollars) of the impacts of the 1986–87 and 1991–95 events *should* be decreasing. Forecasts of El Niño now abound. They are issued by government agencies in the USA, Canada, Australia, and Europe, as well as by consulting firms in various sectors of society and different research groups. Today, societies are somewhat forewarned and, as the saying goes, forewarned *should* mean forearmed. However, this is not always the case.

At present there is too much emphasis in the scientific research community on developing an El Niño forecast capability and, at the same time, an underemphasis on the value of using existing El Niño information. In the meantime (and it could be years, if not decades), the research community must provide society with the El Niño information it has already produced so society can, if possible, use it to adapt to, mitigate, or prevent the worst consequences of future events. No doubt the use of El Niño information has tremendous *potential* value to societies. However, the societal aspects of El Niño have been ignored for too long by funding agencies, even though there are many important issues related to these aspects that need to be addressed, such as who really needs El Niño information.

The forgotten El Niño

Although the 1982–83 El Niño overshadowed all the preceding events, it was the 1972–73 El Niño event that prompted the scientific community to pay increased attention to this phenomenon. At the time, the 1972–73 El Niño event was "almost certainly the most intense observed since 1891" (Cushing, 1982, p. 290). In the early 1970s, for the first time since the end of World War II, global food production declined. Anomalous weather in the early 1970s sparked droughts in the Soviet Union, West Africa, Ethiopia, India, southern Africa, Australia, Central America, Brazil, and Indonesia. Fish landings also declined for the first time since the end of the war. The climate anomalies in the early 1970s sparked the convening of the World Food Conference in Rome in 1974, which was followed by a series of United Nations conferences on other global issues, such as on population (Romania), human settlements (Mexico), water (Argentina), desertification (Kenya), climate (Switzerland), and technology (Austria).

The 1982–83 El Niño captured the attention of the media, turned the phenomenon of El Niño into a household word in the USA, opened the eyes of policymakers to the fact that El Niño events could cripple developing economies by disrupting trade in certain agricultural commodities such as wheat and maize, and reinforced the need for organizing a decade-long research program to improve our understanding of El Niño. The 1972–73 event, however, brought the topic of El Niño to the forefront of the scientific research agenda. The worldwide costs in lives lost and properties damaged as a result of the 1972–73 and 1976–77 El Niño events were not calculated. However, a qualitative expression of the likely worldwide damages associated with that El Niño event is summarized in the climate impacts map shown in Figure 5.2, overleaf.

The 1972–73 event clearly merits a place in the yet-to-be-created El Niño "Hall of Fame," as *the* event that energized the oceanographic, atmospheric, and biological research communities and also prompted some of the first papers on the societal impacts of El Niño.

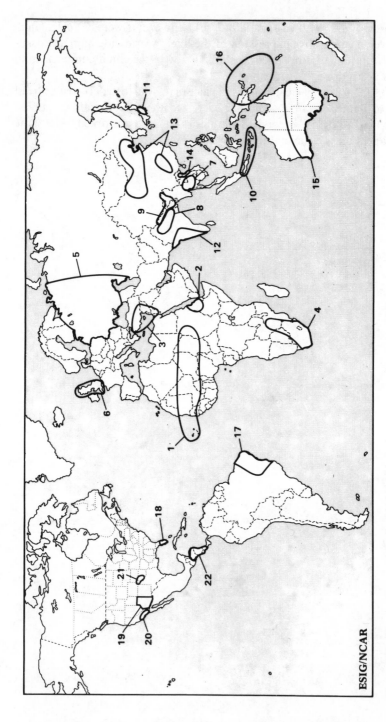

Figure 5.2. Climate impacts map highlighting drought conditions worldwide in March to December 1972.

ESIG/NCAR

Key to Figure 5.2

Africa & Mid East

1. African Sahel: Mali, Mauritania, Senegal, Cape Verde Islands, Niger, Upper Volta, Chad, w Sudan – Mar–Dec
   a. Drought in Sahelian region that began in 1968 intensified in 1972, crop failures, livestock deaths, water shortages, famine
   b. Chad: 1972 annual rainfall lowest since 1943
   c. Agades, Niger: 1972 rainfall total 30 mm vs 164 mm average

2. n and c Ethiopia: Tigre, Wollo, Eritrea, Showa provinces – Mar–Dec
   a. Drought began in 1971, extensive crop losses, 80% of cattle lost, famine

3. Mediterranean Mid East: Jordan Valley, Syria, Turkey, Israel, Lebanon, Cyprus – Sept–Dec
   a. Drought said to be worst of century, water shortages in many urban areas, livestock losses, crop damage, no rainfall in some areas during all of 1972

4. s Africa: Zimbabwe (Rhodesia), Republic of S Africa – Nov–Dec
   a. Failure of summer rains, crop and livestock losses

Europe and USSR

5. USSR: European areas – May–Sept
   a. Worst drought in century: less than 25% of normal rainfall in July and Aug over large area
   b. Mid July–Aug heat wave in ne USSR caused fatalities, crop damage, forest and peat fires (over large areas of e Russia)
      (i) Hottest summer on record in Moscow, heat wave extended into n Finland

6. United Kingdom: Britain – July–Dec
   a. Many locations reported 1972 annual precip. as lowest since 1921
   b. Water shortages in many areas, esp. Scotland and e areas, crop damage
   c. Unusually cold and dry winter also hurt wheat crop, spring also dry

Asia

7. Thailand – April–June
   a. Drought during beginning and middle of rainy season, crop and livestock losses

8. n India: W Bengal, Uttar Pradesh, Bihar – May–early June
   a. Heat wave and drought, temps consistently over 100°F, death toll over 600, New Delhi severely affected damage to crops, livestock

9. ne India, Nepal, Bangladesh – June–Aug
   a. Failure of summer monsoon
   b. Assam province, India received 855 mm rain in July compared to 2855 mm average

10. Indonesia: Java, Madura, Bali and se areas – June–Oct

11. Japan: Tokyo area – June–early July
    a. Drought, worst water shortage in area since 1964

12. w coastal India – July–Dec
    a. Severe monsoon drought in 4 western states along coast, esp. Maharashtra, crop damage

13. China – July–Dec
    a. Widespread drought, crop damage, livestock losses, water shortages

14. n Vietnam: 5 provinces s of Hanoi – Nov–Dec

Australia & Oceania

15. s Australia – Mar–Jan 1973
   a. Severe drought; sw W Australia in 3rd year of dry weather, winter 1972 rains below average for most areas in s Australia, conditions worsened as drought extended into summer rainfall areas of se, drought ended late Jan 1973 and into Feb with heavy rainfall
   b. Affected areas included: sw W Australia, S Australia, New South Wales, s Queensland, s Northern Territories, Victoria (esp. Victoria)
   c. Heat waves
      (i) May 1972: Sydney and neighboring areas, bushfires in New South Wales and Victoria
      (ii) Dec 1972: se areas, esp. Victoria, persistent heat heightened effects of drought
      (iii) 20–23 Jan 1973: Melbourne areas, fatalities
16. Papua New Guinea and Melanesian Islands – April–Aug

USA, Canada, Latin America, & the Caribbean

17. ne Brazil: esp. Ceara, Bahia, Piaui – Mar–Dec
   a. Effects of drought began to be felt in April, severe in some locations by Aug
   b. Area experienced very severe drought in 1970–71
18. se USA: s Florida – Mar–Nov
   a. Reported in Nov that previous 29 months had below-normal precip., crop losses, Everglades "threatened"

19. sw US: Arizona – Mar–May
   a. Jan–April in Phoenix and Tucson completely rainless, driest 4 months on record
   b. Extended drought on Navajo reservation, esp. severe by end of May
20. w USA: s California – Mar–Dec
   a. Los Angeles area rainless from 28 Mar–end of year
   b. San Diego, 1st 4 months of 1972 driest on record
      (i) 1st 4 months of 1972, April in particular, also very dry in other parts of s & w USA: El Paso, 90 consecutive days of no rain reported at end of April; Grand Junction, CO, driest Jan–April on record; Cheyenne, WY, April was 6th consecutive very much dry month; Charleston, SC, driest April on record; Macon, GA, 2nd driest April on record
21. USA: w Oklahoma – Mar–May
   a. Worst drought in w Oklahoma since 1930s, loss of 25% of wheat crop, reported as 2nd year of drought in area
22. Central America: Honduras, Nicaragua, Costa Rica – June–Dec
   a. Abnormally light rains during summer rainy season, crop damage
   b. Drought also reported in El Salvador during June and July
   c. Nicaragua reported severe 1972 crop losses: 80% corn, 35% beans & rice, 20% wheat (reported in Jan 1973)

6   Forecasting El Niño

The value of a forecast: What is versus what ought to be

Different words are used to describe predictions of what might happen in the future. A popular word for this is forecasting. Another might be projection – yet others, scenario, outlook, or prognostication. Sir Gilbert Walker used the term "foreshadow." Each of these words is used in order to present an image of some aspect of the future. The words are also found in the scientific literature, especially in El Niño research and popular literature. Some scientists say that they are not making forecasts but do acknowledge that they are offering projections.

---

fore·cast \ fōr-käst \ *vb* **forecast** *also* **fore·cast·ed; fore·cast·ing** *vt*
(15c)   **1   a:** to calculate or predict (some future event or condition) usu. as a result of study and analysis of available pertinent data; *esp*: to predict (weather conditions) on the basis of correlated meteorological observations   **b:** to indicate as likely to occur   **2:** to serve as a forecast of: PRESAGE ⟨such events may~peace⟩   ~ *vi* : to calculate the future   *syn* see FORETELL – **fore·cast·able**   *adj* – **fore·cast·er** *n*
Merriam Webster's Collegiate Dictionary, 10th edn., 1993

While those concerned with semantics might feel that there is ample distinction between these terms as they are used by scientists, all descriptions attempt to provide to a target audience such as the public, policymakers, or other scientists a glimpse of what particular forecasters think the future climate will be like.

Forecasts are just probability statements; they have a chance of being right. There is also a chance that they will be wrong. While forecasts might prove to be right much of the time, one should also expect them to be wrong sometimes as well. Attempts to assess the value to society of a forecast capability should not be based on the success or failure of any single forecast. The value to society of a forecast (or a forecasting system) can best be measured by assessing over time the success or failure of a series of forecasts. El Niño- or Southern Oscillation-related forecasts should be assessed in this way. Forecasts must be used with care, and the more dependent one becomes on them, the more care that must be taken with their use. Forecasts do not come with warrantees; there are identifiable risks associated with their use.

The following is the opening paragraph in a Queensland (Australia) state booklet on El Niño and the Southern Oscillation.

> "If I'd known that it was going to stay so dry, I would have sold more cattle early in the season," said the grazier. "If I'd known it was going to be so wet, I wouldn't have bought that new irrigation pump," said the farmer. If those whose livelihoods depend on the weather had a better idea of the coming season, they could make better decisions.
>
> (Partridge, 1991, p. 1)

It expresses succinctly the belief of most people that having more information about weather or climate has got to be better than less information. But assessing the value to a farmer, to a grazier, or to a society is not such a simple, straightforward task. As there is no such thing as a perfect weather forecast, forecasts in Queensland are often issued in terms of probability. The booklet, *Will It Rain?* also notes that "We can improve the probabilities based on historical rainfall records. Meteorologists have found that the Southern Oscillation can allow better prediction of rainfall in the coming season." The report continues,

> Monthly rainfall totals can be checked mathematically to show the distribution of wet or dry years and, from this, we can express the chances, or probability, of getting more than a certain amount of rain in any month. But a 66% probability of getting more than, say, 100 mm of rain still means that one year in three we could make a wrong decision.
>
> (Partridge, 1991, p. 1)

The value of reliable predictions of El Niño was recognized by scientists at least a few decades ago. In 1961 Jacob Bjerknes referred to the potential value of forecasting El Niño's onset. He suggested that

> This particular meteorological phenomenon will soon be photographically reported from "Tiros" or (later) "Nimbus" type weather satellites. With such data at hand, it may become possible to discover the first beginnings of a developing "El Niño" at sea early enough to issue useful "El Niño" warnings for the coastal fisheries.
>
> (Bjerknes, 1961, p. 219)

Later, University of Hawaii oceanography professor Klaus Wyrtki and his colleagues also explicitly commented on this:

> The economy of Peru is strongly influenced by its fishery, as is the world market of protein for animal feed, and so a prediction of the occurrence of

El Niño would be a valuable guide for long-range economic planning. ...
[*A*] capability to predict this event would contribute to the understanding
of the large changes occurring in our weather and climate.

(Wyrtki et al., 1976, p. 343)

However, what ought to be the value (in the best of all possible worlds) may
not be what is actually achieved, given the various social, economic, and
political constraints affecting the ideal use of such information at any given
time. These are the two different perspectives on a forecast's value.

If you were to ask a farmer or an industry representative what one might
do with perfect knowledge about a future weather event, he or she would
probably respond with numerous tactical uses of that information. With
regard to anchoveta fisheries off the coast of Peru, for example, some might
argue that the information could be used to protect the fishery against
overexploitation; tie up the fishing boats during El Niño so that they will
not take too many fish from the sea at a time when the fish population is
most vulnerable; shorten the fishing season; limit the fish catches; change
the mesh size of the fishing nets; and so forth.

From the perspective of the fishing sector, however, many of the tactical
measures proposed to be taken in response to an El Niño forecast to protect
the resource would place major financial hardships on the fishermen, the
boat owners, the crews, ancillary industries such as shipbuilding, and banks
that depend on the fishing industry. If fishermen are not allowed to fish,
how can those who are dependent on fishing for a living cover their monthly
payments to the bank for the boats or equipment? The reality is that
fishermen continue to demand the right to fish, even during El Niño. The
only thing between them and their all-out exploitation of the fish is
government decisions to stop all fishing efforts for months at a time (closed
seasons are called *vedas* in Spanish). Fishmeal-processing plants, too, have
constant demand for fish to process into meal and oil. Demands in the
international marketplace for fishmeal also put pressure on the Peruvian
fishing industry to maintain production, especially when fishmeal prices are
high.

Therefore, when the public listens to scientists as they theorize about the
many ways that an El Niño forecast might be of value months in advance of
its onset, it is important to keep in mind that it will be difficult to realize all
that potential.

Going public

One can imagine that a scientific prize will likely go to the scientist in the
El Niño research community who first manages to develop a method for
producing highly reliable forecasts of the *onset* of El Niño several months in

advance. But many researchers are interested mainly in seeking to understand some of the pieces of the El Niño puzzle, such as the oceanographic, atmospheric or biological processes associated with El Niño. The scientific community, however, is not certain that all of the puzzle's pieces are as yet on the table, since new aspects of El Niño processes are continually being uncovered. For the most part, scientists are reluctant to "go public" with their thoughts about whether an El Niño event may be in the offing. While there are researchers who feel uncomfortable about making such public projections, others feel compelled to speak out, because of their interpretations of the existing El Niño information.

As confidence increases in different forecast-producing schemes, more and more scientific groups are likely to go public with their projections about the onset of El Niño events, using whatever indicators they feel are most relevant. Some base their El Niño forecasts on changes in the ocean environment, such as changes in sea surface temperatures or changes in ocean currents. Others tend to base their projections on observed changes in the atmosphere, such as changes in sea level pressure differences across the Pacific basin or surface wind direction. Still others use statistical or analogue methods or computer models to forecast the onset of events.

Does the present state of knowledge about forecasting the onset of an El Niño merit the marketing of forecasts to the public as a "done deal"? Is it a usable forecast, even though its record of reliability over the long term has yet to be established? El Niño forecasters believe that the reliability of El Niño forecasts has already been overwhelmingly established. Recent forecasts of El Niño events provide some examples of forecasts that proved to be correct. There are also examples of missed forecasts. Both kinds of example are provided in the following sections.

---

re·li·abil·i·ty \ ri-'lī-ə-'bi-lə-tē \ n (1816)   **1:** the quality or state of being reliable   **2:** the extent to which an experiment, test, or measuring procedure yields the same results on repeated trials
re·li·able \ ri-'lī-ə-bəl \ adj (1569)   **1:** suitable or fit to be relied on: DEPENDABLE   **2:** giving the same result on successive trials –
re·li·able·ness n – **re·li·ably** \ -blē \ adv
Merriam Webster's Collegiate Dictionary, 10th edn, 1993

Forecast successes

Cane/Zebiak: 1986–87

Perhaps the first notable success story centers on the forecast made by Stephen Zebiak and Mark Cane of the Lamont-Doherty Earth Observatory at Columbia University. In the early months of 1986, these researchers decided to "go public" with their forecast of an impending El Niño for the 1986 and 1987 period. Their forecast was based on the use of a simple oceanic and atmospheric model (called a coupled model) focused on the equatorial Pacific region (Cane *et al.*, 1986). Many of their scientific colleagues expressed concern because of their decision to present their forecast to the public, believing that the research findings of Cane and Zebiak were not yet sufficiently reliable to present. The development of a moderate El Niño in late 1986, however, proved that their projection was correct. This success and perhaps the publicity that accompanied it encouraged other researchers to begin to issue public forecasts of El Niño by the end of the 1980s.

Ethiopia: 1987

Following a major famine in Ethiopia in the mid-1980s, during which hundred of thousands of people perished, the attention of international political leaders focused on the urgent need for a famine-prevention program for the Horn of Africa. One resultant activity was the regional development of early warning systems for food security in general and famine prevention in particular. Another activity was to search for early indicators of changes in rainfall that have been closely associated with crop failures. In this connection, the National Meteorological Services Agency (NMSA) started to issue seasonal forecasts and brought to the attention of concerned high-level authorities the possibility of drought during the 1987 main rainy season.

Therefore, the NMSA was well aware at that time of the apparent linkages of El Niño occurrences thousands of kilometers away in the equatorial Pacific to the failure of rains in Ethiopia's main ("Big") rainy season from June to September. One particular report entitled, *The Impact of El Niño on Ethiopian Weather*, issued in December 1987 (the end of an El Niño year), noted that

> several investigations have revealed that the rain-producing components in Africa are either weakened or displaced or both in ENSO years. ... However, it is not only the ENSO events that cause the recurrence of droughts. Other investigations have indicated that the warm sea surface temperature anomalies over the southern Atlantic and Indian oceans during the rainy season have considerable influence on the recurrence of droughts in Africa. ... The better understanding of the occurrences of ENSO events and their influence on Ethiopian weather are indispensable aids in preparing and disseminating long-range weather prediction.
>
> (NMSA, 1987, pp. 1–2)

By looking back at Ethiopia's yearly climate conditions during both El

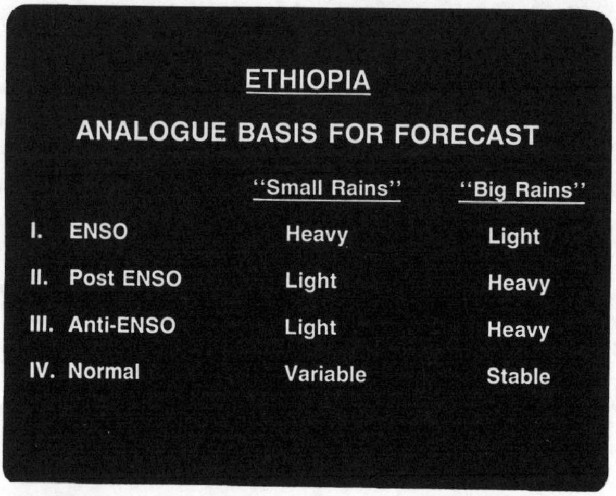

Figure 6.1. Chart produced by Ethiopia's National Meteorological Services Agency that suggests the possible impacts on its "Small Rains" season (mid-February to mid-May) and "Big Rains" season (June to September) of El Niño–Southern Oscillation.

Niño and non-El Niño years and identifying what their seasonal rainfall patterns had been like during these periods, Ethiopian meteorologists were able to suggest with some degree of confidence to government officials what the prospects might be for rainfall, and, therefore, for agricultural production, in the upcoming seasons. This analogue basis for making such forecasts (Figure 6.1) was a popular method among North Americans and Europeans in the middle decades of this century. Today, however, it is not considered a robust (i.e., reliable) approach to seasonal forecasting, unless it is used in combination with other information.

Whenever a forecast of El Niño is issued, someone is likely to take action based on it. A forecast of the onset of El Niño for late 1986 and 1987 was no different. It generated considerable concern about a possible return of famine-like conditions to Ethiopia. Ethiopian meteorologists and government took the prospects of an ENSO event very seriously and sought to modify the "normal" behavior of farmers. The government encouraged farmers to engage in all-out production during the "small rains" (that is, the short rainy season) that occur from mid-February to mid-May, in anticipation of losses that could result from severe drought during the main rainy season. As a result, the forecast of Ethiopian rainfall based on an El Niño forecast enabled the Ethiopian government to take mitigative action. The anticipated severe drought did indeed occur during the season of the "Big

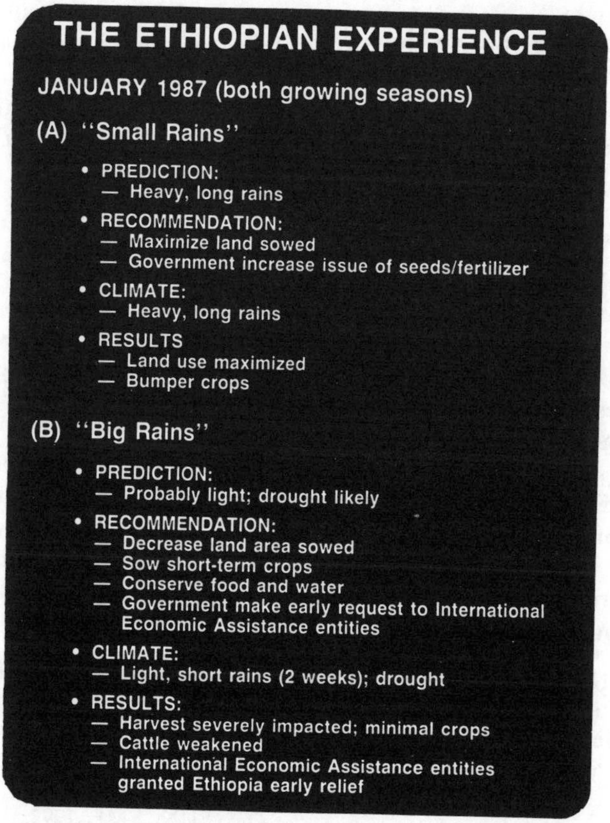

```
┌─────────────────────────────────────────────────────┐
│                                                       │
│  THE ETHIOPIAN EXPERIENCE                             │
│                                                       │
│  JANUARY 1987 (both growing seasons)                  │
│  (A)  "Small Rains"                                   │
│                                                       │
│     •  PREDICTION:                                    │
│        —  Heavy, long rains                           │
│     •  RECOMMENDATION:                                │
│        —  Maximize land sowed                         │
│        —  Government increase issue of seeds/fertilizer│
│     •  CLIMATE:                                        │
│        —  Heavy, long rains                           │
│     •  RESULTS                                        │
│        —  Land use maximized                          │
│        —  Bumper crops                                │
│                                                       │
│  (B)  "Big Rains"                                     │
│                                                       │
│     •  PREDICTION:                                    │
│        —  Probably light; drought likely              │
│     •  RECOMMENDATION:                                │
│        —  Decrease land area sowed                    │
│        —  Sow short-term crops                        │
│        —  Conserve food and water                     │
│        —  Government make early request to International│
│           Economic Assistance entities                │
│     •  CLIMATE:                                        │
│        —  Light, short rains (2 weeks); drought        │
│     •  RESULTS:                                        │
│        —  Harvest severely impacted; minimal crops    │
│        —  Cattle weakened                             │
│        —  International Economic Assistance entities   │
│           granted Ethiopia early relief               │
│                                                       │
└─────────────────────────────────────────────────────┘
```

Figure 6.2. Chart depicting how Ethiopian decisionmakers used the forecast of El Niño in 1986 to maintain national food security.

Rains," and the government's action reduced the amount of food relief that would otherwise have been needed from the international community. Above all, because of the timely action taken by the government, no single human life was lost because of this severe 1987 drought in the region (Figure 6.2).

Cane/Zebiak and the National Meteorological Center: 1991

A few years later, in 1990, researchers at various research centers decided to issue their forecasts of El Niño for that year to the public. Cane and Zebiak, however, did not agree. An El Niño event failed to materialize in 1990, but in the early months of 1991, Cane and Zebiak, using their wind-driven ocean model (Cane *et al.*, 1986), forecast the onset of an El

Niño for late 1991. Once again, they were correct. Forecasters at the United States government's National Meteorological Center, using a statistical model, were also correct with their forecasts. With these two successes to point to, the forecast community became much more encouraged about the prospects of developing a truly usable reliable long-range El Niño forecast.

Northeast Brazil (Nordeste): 1991

Drought in Northeast Brazil is not an uncommon occurrence. In fact, many famous novels have been written about the plight of inhabitants of the region since the 1800s, the most famous of which is Euclides da Cunha's *Rebellion in the Backlands* (Cunha, 1944). In fact the Brazilian parliament has held debates since the mid 1800s about what to do to mitigate the impact of drought in this vulnerable region.

Recent experiences in Northeast Brazil provide another example of a successful El Niño forecast. The governor of the state of Ceará had become convinced by Brazilian and international scientists that El Niño events were associated with recurring severe droughts in the Nordeste region. A climate monitoring bulletin of the Meteorological Foundation of Ceará (FUNCEME) mentioned the linkages as follows:

> The El Niño phenomenon begins to affect the coast of Peru. When this localization of El Niño is finished, its interference in the rainy season takes effect. During March, when El Niño becomes established on the northwest coast of South America, the regular rainfall in the entire Northeast region becomes reduced.
>
> (FUNCEME, 1992, p. 13)

The state of Ceará supports this Foundation, both morally and financially, in its attempts to forecast the probability of regional droughts. In fact, the Foundation had been created to inform government leaders about the climate situation in the region from season to season and from year to year.

A major success for FUNCEME occurred in December 1991, when its forecasters issued a warning of severe drought based on an assessment from the United States National Weather Service of the possibility of the onset of El Niño. The governor, believing the forecasters, traveled throughout Ceará to encourage people to respond to the threat of drought by planting crops that could grow and mature in a drier, shorter-than-usual growing season. In addition, warnings were issued to the people in Fortaleza, the capital city of Ceará, that there would be severe urban water shortages in the event of drought. As a direct response to this warning, rationing was imposed by the government as a precautionary measure. The governor also decided to support the construction of a new dam to "stockpile" water resources.

With a drought in Ceará in 1992, these proactive responses were considered to have been successful, as can be seen by comparing the severity and impacts of the 1992 drought with a similar one in 1987.

Year	Precipitation (% of average)	Grain production (metric tons)	Grain production (% of average)
1987	70	100 000	15
1992	73	530 000	82

The assumption of some observers was that the major difference between the impacts on agriculture of these two droughts was the availability and use of an El Niño forecast for 1991–92. It had apparently influenced the crop production activities of farmers (Lagos and Buizer, 1992).

Australian examples

The high variability of rainfall in northern and eastern Australia has been strongly associated with the regional influence of the El Niño phenomenon (as broadly defined) and especially the Southern Oscillation. The 1982–83 El Niño event visited upon Australia its worst drought. Its most recent prolonged drought was also costly, with adverse impacts mounting to over $AUS3 billion. Although droughts have plagued Australia in the past and their connections to variations in sea level pressure across the Pacific basin have been known for decades, it was the devastation (bush fires, dust storms, agricultural and livestock losses, and so forth) in 1982–83 that prompted the widespread interest in El Niño among the public, policymakers, and the scientific research community in the country.

The Southern Oscillation Index (SOI) has been used in combination with other relevant information to forecast the possible occurrence of good and poor rains. With some lead time, decisionmakers would have options to manage the exploitation of their resources (such as their grazing lands) more efficiently. Graeme Hammer, principal scientist in Queensland, has noted,

> there is potential value of seasonal forecasting to government in considering a range of policy issues (macroeconomics, trade, taxation) as a consequence of the pervasive influence of ENSO on the Australian economy.
>
> (Hammer, 1995, p. 6)

Hammer has been quite active in applying El Niño information to the needs of farmers and graziers at the local level. There have been several studies on the application of El Niño and of Southern Oscillation information in decisionmaking. These have been summarized by Nicholls (1991), a leading proponent of the societal use of ENSO information, especially the SOI. These studies have focused on possible linkages between ENSO and changes in the frequency, location, and intensity of tropical cyclone activity in the eastern Australian region, the SOI and low river flow in the Darling River, droughts and El Niño events, El Niño events and fluctuations in waterfowl numbers in southeast Australia, trends in the SOI and wheat and sorghum yields, and so on. The Australians have also sought to predict outbreaks of mosquito-borne Murray Valley encephalitis in southern Australia using El Niño information (Nicholls, 1986).

One could easily show that Australians have been relatively more active in searching for, identifying, and applying ways to enhance the use of such information to their economic development activities at all levels of decisionmaking. Australians have had their share of successes and problems in forecasting El Niño episodes and, therefore, their impacts. However, as Hammer and Nicholls, among others, have noted, "the value of seasonal forecasting is not a one-off occurrence. . . . The potentially large benefits will accrue over time as ENSO cycles continue to occur."

Missed forecasts: Some examples

Along with success stories such as these, there have been some notable misses with regard to forecasting El Niño or its impacts in the past 20 years. These examples help us to put realistic limits on our expectations about the value of El Niño information to our decisionmaking needs.

Quinn–Wyrtki forecast: 1974–75

Early in 1974, American oceanographers William Quinn and Klaus Wyrtki believed that, on the basis of their research findings about the phenomenon, an El Niño was likely to develop in early 1975. This took place early in the period of research focusing on El Niño processes in the equatorial Pacific region, when less was known about them. Quinn and Wyrtki's projection was supported by other researchers in the fall of 1974. On the strength of those beliefs, the United States National Science Foundation funded, on very short notice, two research cruises to be carried out in the eastern equatorial Pacific between February and May 1975.

Observations taken during the first cruise reinforced the Quinn and Wyrtki conviction about the likelihood of the impending development of an El Niño. The second cruise a few months later, however, yielded observations that the emerging El Niño conditions had transformed into a cold event, a phenomenon in which at that time there was no scientific interest. The field program was then canceled and the ships returned to port. Their forecast of El Niño was considered to have been "blown". Some researchers argued at the time that, although a full El Niño did not develop in 1975, new insights into oceanic processes were gained. For example, although El Niño-like processes may appear in the eastern Pacific they may not emerge as El Niño events (Wyrtki *et al.*, 1976).

Wyrtki's view of their forecast and the "collapsed" El Niño can be summarized as follows:

> The year 1975 will not enter oceanographic history as a year of a large El Niño. However, as predicted, an El Niño situation started to develop with a characteristic overflow of warm, low salinity water from the north, an intensification of the undercurrent, and an accumulation of sub-surface water along the coast. Without the El Niño expedition, these conditions would not have been observed, and from coastal temperatures alone one would have concluded that nothing abnormal had happened. This investigation indicates that only very strong El Niño events have been recorded in the past and that weak occurrences have remained unnoticed.
> (Wyrtki *et al.*, 1976, p. 346)

An El Niño event did develop a year later in 1976. It is only fair to note that modelers who try to reproduce anomalies that have occurred in the past in order to verify their models have typically had difficulties in replicating El Niño events in this period. One researcher suggested that Quinn and Wyrtki were unlucky to have started to forecast El Niño with the toughest case in the past 25 years or so.

The 1982–83 El Niño

As noted earlier, most El Niño researchers, with a few exceptions (Nicholls in Australia and Rasmusson in the USA), failed to forecast the onset of this major event (see Chapter 5).

A 1982–83 El Niño teleconnection forecast

Another notable forecast failure with regard to El Niño's impacts took place in early 1983, when a university professor issued a forecast of the impacts on agriculture in the United States Midwest of the El Niño that began in late 1982. He projected, on the basis of about 100 years of El Niño and of corn yield records, that corn yields in Illinois would be extremely high. His projection was given very high visibility when it was published in the journal *Science*. Based on this article, his projection was reported in major American newspapers. The projection of corn production, however, proved erroneous (i.e., dead wrong) when, later in the season, it became clear that corn yields had actually been reduced by about 50%. This erroneous forecast undermined, for a while, the credibility among scientists of such teleconnection forecasts.

Forecasts in the 1991–95 period

The first half of the 1990s proved to have been a nightmare for the El Niño forecasting community. Things seemed to go well at first. The onset of an El Niño was correctly forecast by the United States National Meteorological Center (NMC) and by Cane and Zebiak in early 1991. However, the demise of the 1991 event expected to take place by the end of 1992 did not occur. A weak El Niño began in the eastern equatorial Pacific again in 1993, died out, only to be followed by yet another event in 1994.

Even when one looks at the sea surface temperature anomalies in the first half of the 1990s for the different regions of the equatorial Pacific (i.e., Niño1 to Niño4; see Figure 4.6), it becomes difficult to gauge the El Niño situation because of conflicting signs. Figure 6.3(*a*) shows sea surface temperature anomalies in Niño1 and Niño2 regions, those regions along the western coast of equatorial South America. One can see three distinct but not very strong El Niño-like warmings in early 1992, in early 1993, and in late 1994.

The next chart (Figure 6.3(*b*)) depicts anomalies for the Niño4 region in the western equatorial Pacific for the same period of time. One can see that the sea surface temperatures stayed above average from late 1989 to early 1995. Thus, researchers monitoring the Niño4 region might argue that the early 1990s were affected by one long El Niño rather than three shorter ones in the same period.

This El Niño has challenged the forecast community because, depending on the region one focuses on, the event could be viewed either as one long situation or as a set of three smaller ones. Yet another view could be that there was an El Niño event in 1991–92 in the eastern equatorial region, followed by the occasional weak, above-average increases in sea surface temperatures in the same region.

The United States National Weather Service (NWS) created a program in early 1995 to produce long-range forecasts. Using observations and a variety of methods, the NWS issues forecasts for each month up to 15 months into the future. Changes in sea surface temperatures in the equatorial Pacific play a central role in these forecasts.

The NWS picked a difficult time to engage in long-range forecasts, in the midst of the unusual behavior of sea surface temperatures across the equatorial Pacific Ocean. As one forecaster noted,

> the El Niño now developing [*December* 1994] came as something of a surprise. ... It is the third in four years, and El Niño forecasts, including

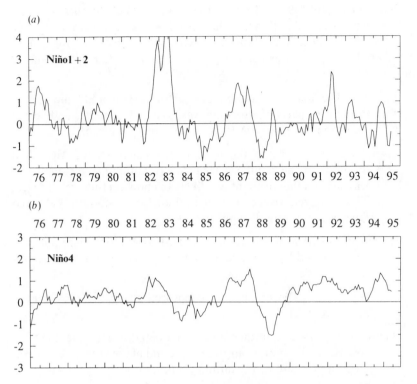

(*a*)

(*b*)

*Figures 6.3. (*a*) and (*b*) Sea surface temperature charts for the 1976 to 1995 period for Niño1 + 2 regions compared to the Niño4 region. The comparison suggests that an El Niño occurred in the western Pacific from 1990 to 1995, while three El Niño events of varying intensities appeared in the eastern Pacific. (From CAC, 1995.)*

that of the NWS's coupled model, didn't see it coming until late summer [1994]. With El Niño so central to extended-range forecasting, that's unsettling some people.

(Kerr, 1994, p. 1941)

Australia's seasonal outlooks in the 1990s: Hits and misses

The decade of the 1990s opened with the promise for Australian scientists to consolidate their routine seasonal outlook system, which began operations in 1989. This system was based on statistical assessments and an understanding of the "phases" of El Niño. The detail of, and confidence in, the system was reinforced by a greatly enhanced international climate monitoring capability and by improvements in modeling the dynamics of ENSO behavior in the equatorial Pacific. That promise was partially fulfilled during the 1991 to 1995 period. Highlights of the successes and failures of the seasonal outlooks in this period were as follows (note that the seasons refer to those in the Southern Hemisphere, which are opposite to Northern Hemisphere seasons: summer is winter, autumn is spring, and so on):

July 1991: The first successful operational forecast of an El Niño-related drought in parts of Australia was issued. The winter and spring conditions in eastern Australia were well predicted.

Austral autumn 1992: On the basis of the usual end of Australian droughts in the Southern Hemisphere autumn, the community was advised that drought was likely to end with this event as well. This, however, was not a useful outlook for New South Wales and Queensland where drought persisted.

Spring 1992 and Summer 1992–93: The SOI rose to near zero and was accompanied by unpredicted extremely wet conditions in southeast Australia.

First half of 1993: Dry conditions reappeared and predictions were in part successful.

Austral spring 1993: Another apparent breakdown in the El Niño episode. Predictions from June to the end of the year were both on target and useful to the farming community.

1994: There was a reemergence of a major El Niño. Predictions from June to the end of the year were both on target and useful to the farming community.

Early 1995: Cautious statements were issued on the probability that the 1994–95 El Niño episode would end. Such statements were well received by the government and the rural sectors. Good rains were received over much of eastern Australia through 1995.

It remains to be seen whether the promise that opened the decade of the 1990s for potentially useful Australian seasonal outlooks will be fulfilled more completely by the end of this decade (Figure 6.4).

Whenever forecasts are issued, officially or unofficially, and appear in either the scientific or popular media, people are listening. Some will take action based on those forecasts, and they will either reap the benefits or suffer the consequences of doing so.

A study of a bad forecast of seasonal water supply in an agricultural region in the North American state of Washington raised an interesting question about the value of long-range forecasting: could the costs incurred because of an erroneous forecast balance out the hypothetical value of a

Bureau of Meteorology Seasonal Climate Outlook – Summary: Dry weather
likely to persist

The value of the Southern Oscillation Index (SOI) for June was about –9, only
slightly above the May reading of –12. The SOI has maintained significant
negative values since March, and is likely to remain negative for the next few
months. Pacific Ocean temperatures are also beginning to show some El
Niño-like characteristics. Periods of low SOI suggest a high probability of
below-normal rainfall during the July to September 1994 period, particularly over
eastern Australia. Recent negative SOI values are a reflection of
higher-than-normal barometric pressure over the northern Australian/western
Pacific region, and lower-than-normal barometric pressure over the central
Pacific, a pattern which has been dominant since 1991.

Overall, the analysis indicates the following for total July to September 1994
rainfall:

The areas most likely to receive below-average rainfall cover most of eastern
Australia, including Tasmania, together with much of South Australia and
southern Western Australia.

Please note: The diagram does not imply that the individual months July,
August and September will follow this rainfall pattern.

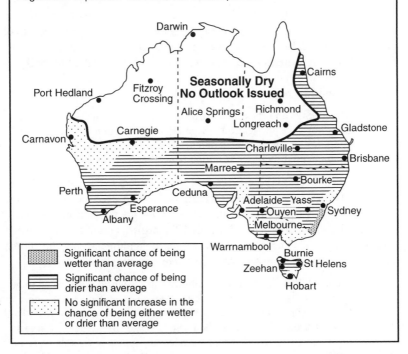

*Figure 6.4. Seasonal climate outlook, July–September 1994. (Issued by the
National Climate Centre, Bureau of Meteorology, Melbourne, Australia, July
1994.)*

larger number of improved forecasts? To assess the value to society of an El Niño-related forecast, one must include the costs of missed forecasts in the calculation of the benefits of correct ones (Glantz, 1982).

The examples of missed forecasts are not presented here to make light of the serious research efforts to develop a reliable and credible long-range forecast system based on an improved understanding of El Niño and the Southern Oscillation. They are provided to underscore the view that forecasting El Niño is a process that is wrought with difficulties. Although progress has been made, a totally reliable El Niño forecast is still not at hand, as witnessed by attempts to forecast El Niño and Southern Oscillation conditions in the first half of the 1990s. However, there is strong pressure from government funding agencies to describe El Niño forecast skills in very positive terms.

While attempts to forecast El Niño are worthy research efforts, deserving of major scientific research funding by governments around the globe, the state of scientific understanding must not be oversold to the public and to policymakers. Scientists should label their forecasts as experimental and should include warnings about the uses as well as the potential misuses of a probability-based El Niño forecast. In fact, recent bulletins of the South African Weather Bureau's Research Group for Statistical Climate Studies (RGSCS) have included an appropriate disclaimer suggesting that the user of the information beware.

> The RGSCS and the South African Weather Bureau accept no responsibility for any application, usage or interpretation of the information contained in this document and disclaim all liability for direct, indirect, or consequential damages resulting from the usage of this bulletin.

Some forecasters have clearly had more success than others. Yet, no forecast group has an unblemished El Niño forecast record. Sir Gilbert Walker issued words of caution about making projections related to weather phenomena in general. In his 1935 address on "Seasonal weather and its prediction," he chose to warn his colleagues about the dangers associated with long-range forecasting. He noted that

> some of the most progressive countries in the world are inclined to make predictions on an insecure basis; their technical staff does not realize that though the prestige of meteorology may be raised for a few years by the issue of seasonal forecasts, the harm done to the science will inevitably outweigh the good if the prophecies are found unreliable... It is the occasional failures of a government department which are remembered.
> (Walker, 1935, p. 117)

Walker's words of the 1930s are still valid today. It was noted above that forecasters in FUNCEME in Northeast Brazil had a few successful forecasts, for which the organization received high praise in international as well as national circles. Most recently, however, following three successful interannual forecasts by FUNCEME, a subsequent seasonal forecast by FUNCEME was perceived by the public to have been in error. As a result, the credibility of the organization, its forecasters, and its forecasts had been sharply eroded. Once credibility has been damaged, it can be a long-term process to get it back. The process to restore the high level of credibility to FUNCEME began almost immediately with the help of other Brazilian research groups.

While FUNCEME took the heat, so to speak, there were other key confounding factors to be considered: these El Nino events of the early 1990s were among the least predictable in recent times. In addition, the previous successes of FUNCEME led the public to feel that the forecasts were going to be absolutely correct. The public did not listen carefully to the caveats (words of caution) associated with the forecast. They came to expect too much perfection from a forecast based on probabilities.

The international research community now believes that it is on the threshold of producing, on a routine basis, reliable operational forecasts of El Niño and of cold events. The United States government's National Oceanic and Atmospheric Administration (NOAA) has taken the initiative to propose the creation of an international research institute for climate prediction. Such an institute would be linked to regional forecast application centers in various parts of the world. While these centers would be dealing with the tailoring of forecasts with various lead times to local and regional conditions, a key concern would be with El Niño- and Southern Oscillation-related, long-lead forecasts.

> The operational prediction of climate fluctuations promises rewards as rich as those of operational weather prediction. A major difference between weather and climate is that the one involves primarily the atmosphere, while the other is a product of the atmosphere interacting with the various water, land and ice surfaces beneath it. Climate prediction requires the concerted efforts of atmospheric scientists, oceanographers, biologists and hydrographers who jointly develop coupled atmosphere-ocean-ice-land models capable of simulating and forecasting climate fluctuations. The activity, because it provides the public with invaluable information, will justify a network of instruments that measure, on a routine basis, not only the atmosphere but also the oceans and land conditions.
>
> George Philander, Princeton University

7 Teleconnections

Teleconnections are among the most intriguing aspects of El Niño. Although linkages between distant climate anomalies were being investigated as early as the mid-1800s, the word "teleconnections" was apparently first used in 1935 by Swedish meteorologist Anders Ångström in a climate research article on the North Atlantic region. Teleconnections can briefly be defined as linkages between climate anomalies at some distance from each other. The large distances in space and the differences in the timing between these anomalous events make it difficult to believe that one could possibly influence the other, but such linkages do exist. There has emerged a subfield of researchers in the atmospheric sciences who focus specifically on improving our understanding of the physical mechanisms behind those alleged teleconnections.

tele- *or* **tel-** *comb form* [NL, fr. Gk *tēle-*, *tēl-*, fr. *tēle* far off]   **1:** distant : at a distance : over a distance ⟨*tele*gram⟩...

con·nec·tion \ kə-'nek-shən \ *n* [L *connexion-*, *connexio*, fr. *conectere*]   **1:** the act of connecting : the state of being connected : as   **a:** causal or logical relation or sequence...

Merriam Webster's Collegiate Dictionary, 10th edn., 1993

In 1975, German scientists Hermann Flohn and Heribert Fleer published a thought-provoking article on climatic teleconnections related to changes in the equatorial part of the Pacific Ocean. They prepared a chart (Figure 7.1) to illustrate alleged teleconnections, suggesting interesting regional climate linkages that, when strung together, encircle the globe. The thickness of the lines suggests multiple-year anomalies as well as increased severity in terms of that particular climate anomaly. Note that the authors separated anomalous climatic episodes in the central Pacific from those along the coasts of Ecuador and Peru, a distinction that became more pronounced in the technical literature in the 1980s than it was in the 1960s and 1970s. According to their preliminary survey, there were only a few

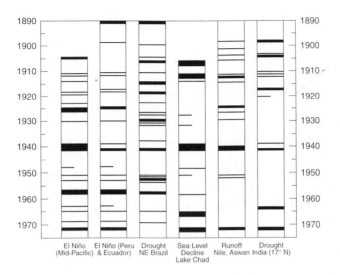

Figure 7.1. Climatic anomalies in six tropical locations; a preliminary survey conducted by Flohn and Fleer (1975), which can be seen by reading the chart from left to right. Thicker lines indicate multiple-year anomalies; half-lines indicate those lasting less than a year. The purpose of the chart was to highlight possible teleconnections between adjacent regions.

occasions (e.g., 1905, 1913, 1923) when anomalous climatic behavior in the central Pacific was not accompanied by similar changes along the western coast of South America. Perhaps in these years the geographic location of changes in sea surface temperatures were somewhat like the situation that appeared in the first half of the 1990s: a warming in the central equatorial Pacific without an accompanying warming in the eastern part.

Flohn and Fleer then linked El Niño with droughts in Northeast Brazil, and droughts in Northeast Brazil to changes (declines) in the level of Lake Chad situated in the midst of the Sahara Desert. Streamflow runoff of the Nile River at Aswan was associated westward with changes in the level of Lake Chad and eastward with droughts in India. As we now know, there have been many attempts to link the droughts in Australia to the failure of the monsoons on the Indian subcontinent, and these droughts have been linked directly to seesaw-like changes in the Southern Oscillation. With the linkages between the Southern Oscillation and changes in sea surface temperatures in the central Pacific Ocean, the chain of anomalies made up of regionally adjacent teleconnections encircles the globe.

Another ENSO teleconnection is associated with temperature over the entire tropical belt. It is observed that, a few months following the peak in the warming of sea surface temperatures during El Niño, temperatures increase over most of the tropics by a degree or so. Likewise, cold events are accompanied by cooler tropical temperatures. These temperature swings are large enough to significantly affect the global average temperature; they must be taken into account when trying to detect global warming of the atmosphere.

Teleconnections, however, are best known with regard to changes in the sea surface temperature in the central and eastern equatorial Pacific Ocean. Does the occurrence of a very strong El Niño move to varying degrees the locations and strengths of the interacting atmospheric high- and low-pressure systems around the tropics and extra-tropics? Does a cold event cause the atmospheric circulation to respond in a sense opposite to that observed in the warm phase? Are there lags in time and/or in space that can be used to forecast whether and when those climate anomalies might occur elsewhere in the future?

Research interest in El Niño during the past two decades has stemmed from a scientific curiosity to better understand physical processes that underlie the Earth's climate. However, as mentioned earlier, interest in El Niño-related research also stemmed from a desire to protect guano production, to protect the Peruvian anchoveta fishery and fishmeal industry, and most recently to forecast climate anomalies within and outside the tropics. The El Niño researchers in North America have become increasingly concerned with the impacts of El Niño events on North American climate anomalies as they affect agricultural production, water resources, commerce, and public safety. The schematic map shown in Figure 7.2 highlights rainfall and temperature patterns that tend (generally but not always) to accompany El Niño events.

It is not yet clear exactly which of the worldwide climate anomalies are associated with El Niño events and which result from "normal" atmospheric processes. Some anomalies give the appearance of being linked to El Niño but happen only by coincidence with sea surface temperature increases in the equatorial Pacific. This raises the problem of attribution: which anomalies can be attributed with a high degree of confidence to El Niño? Some of the apparent teleconnections, however, do appear to be more robust than others and can therefore be used by decisionmakers in a precautionary way.

Recall that, in the Walker Circulation, the rising motion of air has been heated by increases in sea surface temperatures. Where the water warms, ascending motion and the increased likelihood of rainfall can be found. The warmed air mass is carried to high altitudes toward the eastern part of the Pacific basin, where it begins to descend. The rising motion generates the formation of rain-bearing clouds, while the descending motion tends to suppress it. This circulation helps to generate circulation changes at higher levels of the atmosphere.

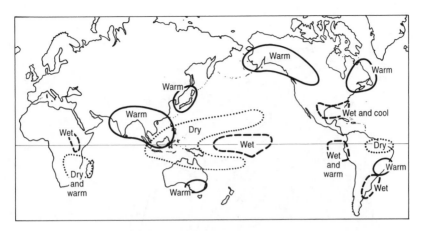

Figure 7.2. Typical rainfall and temperature patterns associated with El
Niño–Southern Oscillation conditions for the Northern Hemisphere winter season.
(From Ropelewski, 1992. Reprinted with permission from Nature, *356,* 476–7.
Copyright 1992 Macmillan Magazines Limited.)

The Walker Circulation is from east to west at the earth's surface and
from west to east at high altitudes. There is also a similar atmospheric
circulation system in the north–south direction, in which air rises near the
equator and descends at the latitudes of the subtropics. The Hadley
Circulation is a direct connection between the tropical atmosphere and
higher latitudes as shown in Figure 7.3. During El Niño, when thunderstorm activities in the Pacific shift eastward along the equator, following the
eastward shift in warm sea surface temperature into the central Pacific, the
jet stream, a belt of strong westerly winds, and its associated storm track
also shift eastward, giving rise to many of the familiar teleconnection
patterns associated with El Niño (Figure 7.2). For example, storms that
usually track northward toward Alaska are shunted eastward toward the
west coast of North America, at times causing flooding in regions such as
southern California, as in the 1982–83 winter. Thus, El Niño episodes affect
both the Walker and the Hadley Circulation patterns, thereby propagating
teleconnections outside the tropical region as well as within it.

El Niño teleconnections outside the tropics are more difficult to pin
down, as they are farther downstream from El Niño's field of action. They
are also difficult to identify because of the numerous possible outcomes of
interaction between El Niño-generated perturbations in the Pacific region
and the regional and local climate conditions in distant regions. *Science*
journalist Richard Kerr commented on the problem, "Although the
equatorial teleconnection appears to be real enough, its usefulness in long
range prediction will be limited to those occasions when the signal from the

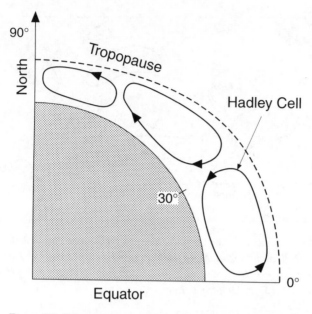

Figure 7.3. Schematic cross-section of the earth, showing equator-to-pole circulation in the Northern Hemisphere. (After Petterssen, 1969.)

tropics rises above other influences on North American weather" (Kerr, 1982, p. 609). Many teleconnections occur simultaneously with El Niño episodes. Those whose occurrences follow the onset of El Niño provide lead time to decisionmakers to prepare to deal with its consequences. Lead time provides value to the El Niño forecast. The climate impacts maps shown in Figures 7.4 to 7.7 provide some examples of worldwide teleconnections associated with the 1991–92 El Niño event.

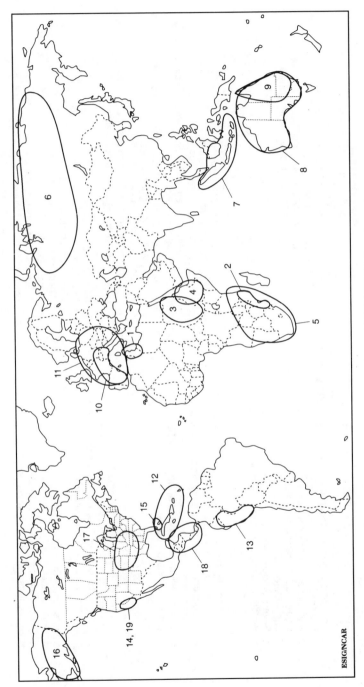

Figure 7.4. 1990–91 droughts. (Compiled by M. Betsill and P. Parisi, ESIG/NCAR.)

ESIG/NCAR

Key to Figure 7.4

Africa & Middle East

1. Tunisia – 1990
   a. Continued drought since 1987; needed to import food
2. Mozambique – 1990
   a. Drought and civil war displaced 1 million people; danger of famine
   b. Agricultural production affected by severe drought in food-producing region of Zambezia
3. Sudan – March 1991
   a. Severe drought; severe hunger in civil war zones
   b. Hundreds of thousands of deaths predicted by relief agencies
4. Ethiopia – July 1991
   a. 7 million Ethiopians affected by drought and famine
5. Southern Africa – late 1991
   a. Drought devastated agricultural production in all Southern African countries
   b. Food situation exacerbated by armed conflicts in Mozambique and Angola

Asia

6. Siberia – late July 1991
   a. Heatwave; temperatures in mid-90s °F
7. Indonesia – late Oct–Nov 1991
   a. Drought; large forest fires in Borneo and Sumatra
   b. Lower production of rice due to drought

Australia & Oceania

8. Australia – April 1991
   a. Drought killed grass; caused bushfires
   b. Sheep farmers faced disastrous year

13. Peru – August 1990
    a. Worst drought this century
    b. Limass river dried up; intestinal diseases increased 52%
14. California – Feb 1991
    a. Drought; severe effects on agriculture
15. Florida Keys – late July 1991
    a. Long Key (65 mi east of Key West); several species of coral began to blanch as water temperature climbed to 31°C
16. Alaska – late July 1991
    a. Heatwave; hotter in Fairbanks, Alaska than in Miami Beach, Florida
17. Midwestern USA – July 1991
    a. Crops wilted by drought
    b. Little or no rain since June
18. Central America – Sept 1991
    a. Greatly reduced rainfall during rainy season; severe drought in region
    b. Reservoirs very low; substantial agricultural losses as a result of poor grain harvests
19. Oakland, California – Oct 1991
    a. Drought-related firestorm destroyed hundreds of homes; $US1.7 billion in insurance claims

Sources

New York Times
The Times (London)
Weatherwise
Climate Monitor
Keesing's Contemporary Archives
Christian Science Monitor
UNDHA Natural Disasters Database
Lexis/Nexus

9. Australia – August 1991
   a. Severe drought in Queensland, New South Wales, and southwestern Australia; rainfall in northern New South Wales and southern Queensland totaled about 1 cm from late July to late October
   b. Worst harvest in 20 years; reduced by 30%; Australia had to import wheat

Europe
10. Southern Europe – summer 1990
    a. Severe drought; France hardest hit
    b. Almost 300 km of rivers dried up in France
    c. Corn output greatly reduced; almost all winter crops were hit

11. Europe – summer 1991
    a. Agricultural outputs in all parts of Europe greatly reduced

USA, Canada, Latin America & Caribbean
12. Caribbean – August 1990
    a. Sea warmed to between 31–32°C for 18 days near Lee Stocking Island; a week later the corals had turned white

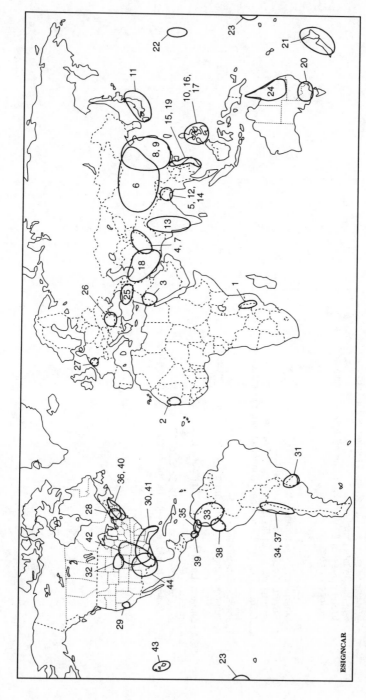

Figure 7.5. 1991 severe storms. (Compiled by M. Betsill and P. Parisi, ESIG/NCAR.)

ESIG/NCAR

Key to Figure 7.5

Africa & Middle East
1. Southeast Malawi – early April
   a. Flooding and mudslides; 516 people dead or missing; 40 000–50 000 homeless
2. Nouakchott, Mauritania – 2 June
   a. Violent winds killed 4 people and injured 300
3. Israel and Jordan – December
   a. Severe rainstorms; major highways blocked and power lines were down; 3 people killed in Israel when a bridge was swept away

Asia
4. Afghanistan – 12 April
   a. Floods; 42 people killed
5. Bangladesh – 27–29 April
   a. Tidal wave submerged many offshore islands, killing 140 000 people
   b. 1 million cattle killed; 1.4 million houses destroyed
   c. $US1.5 bn in damage; hundreds of kilometers of roads and boats washed away
   d. 30 000 people on island of Sandwip swept away; 20 000 died
6. China – 18 May
   a. Floods affected 206 million people; more than 1500 killed and 4 million homeless
   b. $US7.5 bn in damage
7. Northern Afghanistan – early June
   a. Flash floods killed 5000 people
8. Eastern and central China – 9 June
   a. Flooding killed more than 800 people; heavy damage to industrial centers and crops
   b. Nanjing seriously threatened by rising waters of Yangtze River
9. China – mid July–August
   a. 18 July: 200 million people affected by severe flooding
   b. Rainbelt shifted 300 km northward; caused deadly torrential rains and floods near middle and lower reaches of the Yangtze and Huaiche Rivers
   c. By late August, 2295 people had died; 50 000 injured; flooding in 20 provinces
10. Philippines – 12 June
    a. Mt Pinatubo erupted; coincided with typhoon and rainy season
    b. Mudslides and flooding caused deaths of 581 people; 1 million homeless
11. Japan – July
    a. Typhoon Mireille caused damage of $US4.8 bn
12. Bangladesh – 23 July
    a. Floods; 1.5 million people affected
13. Western India – 31 July
    a. Torrential rains; dike broke and river swept into remote village of Mohad; at least 500 people killed; thousands of homes destroyed
    b. In other parts of India, at least 59 people killed in floods or house collapses
14. Northern Bangladesh – 18 Sept
    a. Floods; 50 000 people stranded when a flood barrier broke; 20 million people affected; 500 miles of roads washed away; $US150 million in crop damage
15. Vietnam – 18 Sept
    a. Floods; 21 people killed
16. Northern Luzon, Philippines – 28 Oct
    a. Typhoon Ruth; flash floods; landslides
    b. More than 60 people killed; 21 000 homes destroyed
17. Leyte Island, Philippines – 4–5 Nov

Key to Figure 7.5 (cont.)

a. Tropical Storm Thelma hit city of Ormoc; 4990 people killed; 1389 missing but presumed dead
18. Iran – 17 Dec
  a. Floods; 22 people killed
  b. Fiercest winter storms in over 50 years
19. Vietnam – 28 Dec
  a. Severe storm killed 250 people

Australia & Oceania
20. Australia – June
  a. Wettest June on record in Melbourne
21. New Zealand – June
  a. Coldest June since 1976; reports of sheep frozen to the ground
22. Marshall Islands – 28 Nov
  a. Typhoon Zelda; 11 000 people left homeless
23. Western Samoa and American Samoa – December
  a. Cyclone Val; 13 people killed
24. Eastern Australia, coastal Queensland – late December
  a. 125 cm of rain in 10 days; homes, highways, and railways flooded
  b. 2000 head of cattle lost

Europe
25. Eastern Turkey – 17 May
  a. Flash floods and torrential rain; at least 30 people killed
26. Romania – 30–31 July
  a. In northeast, dam burst after heavy rain; wall of water killed at least 66 people and forced 10 000 people to flee homes

34. Chile – 18 June
  a. Heavy rains; water tank ruptured setting off a mudslide on hillside slums above port of Antofagasta; 116 people killed; more than 17 000 people left homeless
35. Pacific Coast of Panama – July
  a. Two species of reef-building corals went extinct from sea warming and coral bleaching
36. Northeastern USA – July
  a. Hurricane Bob killed 18 people; $US1.8 billion in damage
37. Chile – July
  a. Snow fell in Chile's northern Atacama desert
38. Ecuador – July
  a. Torrential rains in eastern Ecuador destroyed crops
39. Costa Rica – 11 Aug
  a. Floods; 33 000 people homeless; $US15 million in crop damage
40. Northeastern USA and Mid-Atlantic states – 1 Nov
  a. Storms and flooding; smashed homes, boats, sea walls; heavy property damage; 4 people died
  b. Hundreds of millions of dollars in damage
41. Gulf states, USA – winter
  a. Flood in Texas
  b. Along Gulf of Mexico, from Texas to Florida, some areas had 200–1000% of their normal rainfall
42. Midwest USA – Oct 1991–Feb 1992
  a. Iowa had 175% of its normal moisture; Oklahoma 176% of normal; west Texas 255% of normal
43. Kauai, Hawaii – 16 Dec
  a. Storm brought >38 cm of rain in 24 hours, causing flash floods that killed 3

b. In Bucharest, 2 days of flooding claimed 53 lives; 65 people missing; 20 000 people homeless
27. Lint, Belgium – 4 Nov
   a. Heavy rain; community center collapsed killing 3 people, seriously injuring 6

USA, Canada, Latin America & Caribbean
28. New York – 5 March
   a. Worst ice storm in decades; knocked out power to 300 000 homes
29. Stockton, California – 29 March
   a. Flooded Calaveras River swept away 5 young boys; 2 drowned, 2 missing
30. Southern USA – 30 March
   a. Violent thunderstorms and tornados killed 23 people from eastern Texas to South Carolina
31. Uruguay – 16 April
   a. Floods; 1500 people left homeless
32. Crawford, Nebraska – 13 May
   a. Flash floods; one person killed; hundreds of cattle dead
33. Colombia – 4 June
   a. Landslides killed 45 people

44. Texas – late December
   a. North and central Texas: a week of rain; flooding; 15 people killed; ranches, farmlands deluged; cattle and livestock drowned; extensive damage in 28 counties; damage estimated at 10s of millions of $US
   b. 133 families homeless where Trinity River flooded
   c. Southeast Texas: the Brazos River flooded homes, roads, farmland

Sources
New York Times
The Times (London)
Weatherwise
Climate Monitor
Keesing's Contemporary Archives
Christian Science Monitor
UNDHA Natural Disasters Database
Lexis/Nexus

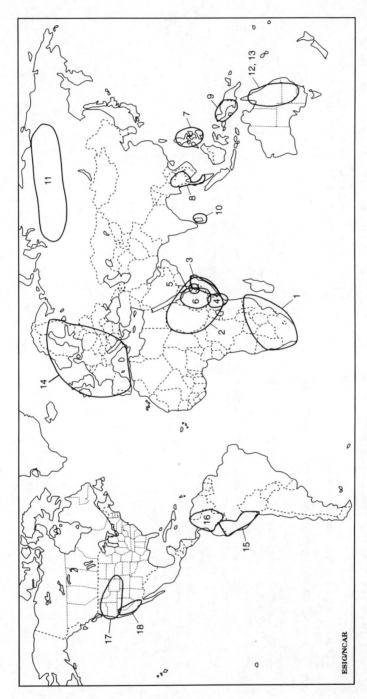

Figure 7.6. 1992 droughts. (Compiled by M. Betsill and P. Parisi, ESIG/NCAR.)

ESIG/NCAR

Key to Figure 7.6

Africa & Middle East

1. Southern Africa – early 1992
   a. Worst drought of 20th century; 80% of commercial corn crop lost; Zimbabwe hardest hit
   b. South Africa and Zimbabwe forced to import large amounts of grain (they are usually grain exporters)
   c. April: 30 million people at risk of malnutrition due to severe drought
   d. Region needed to import 7 million tonnes of food compared to normal imports of 2 million tonnes
   b. Nationwide rationing cost the economy an estimated $US330 000 per week

2. East Africa and Horn of Africa – early 1992
   a. Rainfall less than 50% of normal during first half of year in northern Kenya, southern Ethiopia, and parts of Somalia
   b. Drastic fall in food production; late start of long rainy season
   c. Kenya's dams at lowest levels in 50 years; had to import grain for first time in 8 years
   d. April: millions of livestock dead; crops were not growing
   e. Thousands of people died of starvation (as well as from civil war) in Ethiopia, Sudan, and Somalia
   f. Failure of Meher rains in Ethiopia

3. Somalia – 1992
   a. Food crisis and drought; exacerbated by civil war
   b. 4.5 million people were near starvation; 1.5 million people near death
   c. 500–1000 people died per day due to starvation

4. Kenya–June
   a. 20 months of drought in northeast, east, coast, and Rift Valley provinces
   b. 680 000 Kenyans faced starvation

5. Djibouti – August
   a. 200 000 civilians died at rate of 100 per day due to drought and civil war

6. Ethiopia – October
   a. 200 people died of starvation each week in the Ogaden in remote southeast Ethiopia

Asia

7. Philippines – Jan–May
   a. Drought; $US121 million in crops destroyed; persistent power shortages

8. Thailand – Jan–May
   a. Drought; 12.5 million hectares of farmland destroyed

9. Irian Jaya, Indonesia – February
   a. Lack of crop planting
   b. 132 people died of starvation

10. Sri Lanka – Feb – April
   a. Drought from failure of inter-monsoonal rains
   b. 400 000 of Sri Lanka's 17 million people affected by the drought; drinking water supplies were drying up; tea production down; paddy rice crops affected

11. Siberia – August
   a. Wildfires in southern Siberia; worst drought in 100 years
   b. Crops and large areas of forest destroyed

Australia & Oceania

12. Australia – February
   a. Monsoon rains through eastern New South Wales and Queensland broke year-long drought
   b. Western part of New South Wales remained dry; parts of Queensland also remained dry, affecting sugarcane crops

Key to Figure 7.6 (cont.)

13. Eastern Australia – September
    a. Drought; severe food and water shortages; livestock and wildlife threatened

Europe

14. Europe – late Nov 1991–March 1992
    a. Precipitation less than 50% of normal in southern England, France, Spain, Portugal, and northern Italy
    b. Rainfall less than 25% of normal in Denmark and southern Sweden
    c. Rainfall less than 50% of normal in eastern Norway, most of Sweden and Finland, northern Germany and southwest Poland
    d. Crop production fell in eastern Germany, Norway, and Finland
    e. Drought spread to Baltics
    f. Worst drought in Europe in 100 years; worst in southeastern England since 18th century

USA, Canada, Latin America & Caribbean

15. Andes (Peru) – April
    a. Severe drought; 67 000 alpacas on the brink of death
    b. Worst drought this century

16. Colombia – April
    a. Drought; lack of hydroelectric power and rationing of electricity

17. Western USA – October
    a. Extreme drought from southern Wyoming to Oregon and Washington

18. California – November
    a. By end of Nov, California's 155 largest reservoirs held only 55% of their capacity; one-year loss of 326 bn gallons (1234 bn liters)
    b. Snowpack in Sierra Nevadas 30% of normal

Sources
New York Times
The Times (London)
Weatherwise
Climate Monitor
Keesing's Contemporary Archives
Christian Science Monitor
UNDHA Natural Disasters Database
Lexis/Nexus

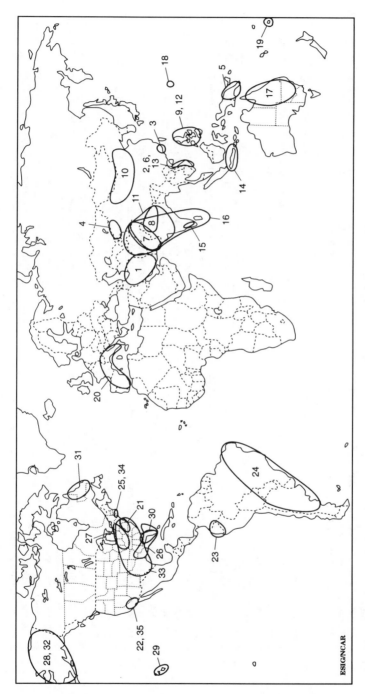

Figure 7.7. 1992 severe storms. (Compiled by M. Betsill and P. Parisi, ESIG/NCAR.)

ESIG/NCAR

Key to Figure 7.7

Africa & Middle East

1. Iran – May
   a. Floods throughout the country; over $US59 million in damage; 42 people killed
   b. 20 villages (40 000 people) cut off near Caspian Sea

Asia

2. Vietnam – 1 Jan
   a. High winds devastated coast, killing at least 100 people

3. Hong Kong – Feb–April
   a. Rainfall more than 3 times that of normal
   b. Highest rainfall since observations began in 1884

4. Kyrgyzstan – May
   a. Torrential rains; more than 5500 houses completely destroyed and more than 4000 damaged; came a week after an earthquake measuring 7 on the Richter scale

5. Papua New Guinea – May
   a. Floods

6. Northern Vietnam – July
   a. Typhoon Chuck; 21 people killed; winds up 120 km/h

7. ᴧakistan – early August
   a. Heavy rains and flooding; 94 dead; 270 000 homes damaged
   b. Storm on 12 Aug in Karachi brought 144 mm of rain in 11 hours, more than half the city's normal rainfall for a year

8. India (esp. northern Himalayas) – early August
   a. Torrential rains and flooding in northern India, especially Kashmir; 980 deaths

9. Philippines (esp. west coast of Luzon) – 15 Aug–5 Sept
   a. Cyclones Mark, Omar, and Polly
   b. 500 mm of rainfall
   c. Flooding forced thousands of villagers to flee

Australia & Oceania

17. Australia – February
    a. Monsoon-type rains in New South Wales and Queensland; widespread flooding in New South Wales

18. Guam – 28 Aug
    a. Typhoon Omar; 248 km/h winds
    b. 200 people injured; $US487 million in damage

19. Fiji – December
    a. Cyclone Joni; wind gusts up to 208 km/h; widespread flooding in coastal villages

Europe

20. Southern Europe – 24 Sept
    a. Torrential rains, violent winds, flash floods destroyed crops and damaged roads and buildings in Vaucluse region of France
    b. Flooding from Mediterranean coast to the Rhone Valley in Provence, France; 80 people died
    c. Flooding in Spain and Italy killed several people

USA, Canada, Latin America & Caribbean

21. Eastern USA – January
    a. Flooding on Assateague, a barrier island outside of Maryland and Virginia, killing 5 ponies that roam there

22. Southern California – 12–13 Feb
    a. Storm, flooding, rain closed freeways, washed out canyon roads, flooded trailer parks; 7 people died; worst flooding in more than 20 years

23. Coastal Ecuador – Feb–June
    a. Heavy rains and flooding; evacuation of 100 000 families

24. Brazil, Paraguay, Argentina, Uruguay – Feb–June
    a. Flooding in all these countries

10. Northern China – 21 Aug
   a. Heavy rain, hailstorms, floods, landslides; at least 111 people killed
11. Pakistan, India, and Afghanistan – early Sept
   a. Northern Pakistan: torrential rains, river flooding caused extensive damage to crops and property; at least 1184 people killed; 2.8 million homeless
   b. 4 million people homeless in Islamabad
   c. Afghanistan: flooding; wall of water, mud, rock swept through valleys of the Hindu Kush mountains; several hundred people died
   d. Northern India: 4 million people displaced
   e. Worst flooding in decades; over 4000 people killed
12. Manila, Philippines – 17 Sept
   a. At least 10 people died in flash floods and mudslides
13. Central Vietnam – October
   a. Typhoon Angela
   b. Flooding; 50 people dead or missing; 28 000 families homeless
14. Java, Indonesia – October
   a. Floods; 77 casualties; 8000 people evacuated
15. India – October
   a. Flash floods and landslides in southern Kerala state; more than 50 casualties
16. India, Sri Lanka, Pakistan – 12–14 Nov
   a. Tidal waves, torrential rains, flooding; 260 mm rain in southern India caused landslides and floods that killed 4000 people; thousands homeless
   b. 13 people died in Sri Lanka
   c. Damage to Pakistan $US2 bn
   d. 1.7 million hectares of agricultural land in Punjab inundated

   b. In early June, heavy rains for 2 weeks caused flooding; 28 people killed; 220 000 forced to be evacuated from their homes in Paraguay, Argentina, and southern Brazil
25. Connecticut – 7 June
   a. Heavy rains caused flooding; 250 people evacuated from homes, 15–20 cm of rain in 15 hours
26. Florida, Louisiana, and Bahamas – August
   a. Hurricane Andrew caused damages of $US16.5 bn
   b. Left 250 000 homeless in Bahamas and Florida; 23 people killed
27. Kentucky, Indiana, Ohio, and western Pennsylvania – 10 Aug
   a. Thunderstorms and flooding; at least 2 people killed
28. Alaska – September
   a. Coldest Sept on record; interior temperatures 10–15 °F below normal; heaviest snowfall in central interior
29. Kauai, Hawaii – 8–11 Sept
   a. Hurricane Iniki damaged 10 000 homes and 70 hotels
   b. Losses around $US1 bn
30. Florida – 3 Oct
   a. Tornadoes at Tampa–St Petersburg; 3 people killed
   b. More than 500 dwellings damaged
31. Newfoundland, Canada – 6–8 Oct
   a. Storm, 80 mm rain, 20 cm snow, 140 km/h winds ripped down trees and utility poles
   b. Damage estimated at over $US9 million
32. Alaska – October
   a. Heavy flooding in Nome
33. USA – 21 Nov
   a. 93 tornadoes in 11 states from Texas to Ohio; 25 deaths
   b. School bus in North Carolina hurled 75 yards

Key to Figure 7.7 (cont.)

34. New York metro area – 12 Dec
    a. Storm with 144 km/h winds destroyed transportation routes and caused floods; knocked out power to hundreds of thousands of homes; damaged thousands of buildings; stretched 960 km along Atlantic seaboard from upstate New York to North Carolina
    b. 20 towns along New Jersey coast evacuated; 6 deaths because of storm in NY; 9 deaths in Pennsylvania
    c. Called Great Nor'easter of '92
    d. 100 beach homes swept away on Fire Island
    e. Insured losses from storm estimated at $US650 million
35. Los Angeles (Orange County) – winter
    a. Heavy storms destroyed homes, buried highways

Sources
New York Times
The Times (London)
Weatherwise
Climate Monitor
Keesing's Contemporary Archives
Christian Science Monitor
UNDHA Natural Disasters Database
Lexis/Nexus

8 Methods used to identify El Niño

There are several ways that scientists try to gather information about El Niño and about cold events to be used to identify El Niño's precursors, to forecast their onset, to monitor their growth and decay, and to identify their teleconnections in the tropics and elsewhere. Some methods have focused on direct observations of characteristics of the warm event/cold event cycle, others on indirect measurements. Still others have used proxy information, i.e., the observation of climate-related phenomena that are perceived to be linked to these events, such as floods, droughts, or bush and forest fires. Each of these natural hazards leaves its lasting mark on the environment.

In addition to observations and direct *in situ* measurements, other assessment methods include remote sensing from satellites, statistical research, and computer modeling activities and simulations. While the results of the various approaches may not agree with one another, for the most part each scientific study, and each new statistical method of assessment, and each new monitoring scheme add to our understanding of the El Niño phenomenon. Even negative findings can be viewed in a positive light because they identify "blind" research alleys.

Before the age of satellites and computer models, researchers interested in the Pacific Ocean (or any other ocean, for that matter) had no choice but to rely for the most part on ships of opportunity (that is, commercial ships that criss-cross various parts of the world's oceans). These ships supply *in situ* information on ocean currents, winds, sea surface temperatures, and so forth. But the Pacific Ocean is a large body of water, and reliance on opportunistic observations of ocean conditions left large gaps in space and time for information on a variety of characteristics related to El Niño.

Statistical methods

Statistics plays a major role in nearly all aspects of research on El Niño and its teleconnections, given that relatively few El Niño events have been directly and knowingly experienced by the recent generation of researchers.

It was the use of statistical methods (such as correlation and multiple regression), pioneered by Sir Gilbert Walker in the earliest decades of the twentieth century, that helped to identify many of the teleconnections associated with the occurrence of the Southern Oscillation phenomenon. Walker also relied on statistical techniques (especially time-series analysis) to identify and model the quasi-periodic features of the Southern Oscillation. The use of statistical methods enabled modelers to incorporate all kinds of relevant physical processes into the modeling and forecasting of El Niño. Current long-lead forecasts of climate anomalies (e.g., issued by the United States National Weather Service Climate Prediction Center) combine output from coupled atmosphere–ocean general circulation models with multivariate statistical techniques (such as canonical correlation analysis or principal component analysis).

The importance of statistics in El Niño-related research cannot be overstated. Statistical assessments pervade all aspects of El Niño research from reconstructing paleoecological records to identifying potential teleconnections to projecting the possible implications for various characteristics of El Niño of a global warming of the atmosphere. Statistical assessments have been a necessary as well as integral part of using El Niño information, for instance, in developing a mechanism for forecasting hurricane frequency on a seasonal basis (e.g., through use of contingency tables).

For their part, social scientists, along with other potential users of El Niño information such as agronomists, hydrologists, and fisheries managers, are also very dependent on the appropriate use of statistics to identify the precise locations and strengths of El Niño episodes so that they can more effectively adapt to, mitigate, or prevent their potential consequences for human activities.

In history

The public tends to think that much of our knowledge about El Niño has emerged only in the last decade. However, as noted earlier, interest in, and research on, aspects of El Niño are not just of recent origin, and can easily be traced back at least to the late 1800s. It was the major El Niño in 1891 that sparked local discussion among Peruvians about El Niño as a phenomenon that should concern them. Following that event, Peruvian navy captain Camilo Carrillo described the history of scientific interest in Peru's coastal waters, going back to the late 1700s and 1800s. His knowledge about El Niño suggests that Peruvians had a great deal of information in hand by the end of the nineteenth century. He spoke of changes in coastal winds and in biological productivity, along with awareness of excessive rainfall in northern Peru and disease outbreaks associated with what we have come to know as El Niño-related changes in ocean environment. So today, much of the El Niño information related to Peru that scientists are using as their base must have been known publicly for about a hundred years.

Anecdotal evidence

Anecdotal information can appear in a variety of ways, such as personal diaries and historical interpretations of past events. For example, in 1895 an American geologist, Alfred Sears, gave a lecture to the American Geographical Society on "The coastal desert of Peru." In addition to a brief description of the flora, fauna, and geology of the coastal region, he referred to the "septennial rains" that occur in the usually arid northern Peru.

These rains seem to have recurred every seven years or so and have since been viewed with good reason as the rains that accompany El Niño events. Sears reported on the great changes that he observed in the desert landscape: "the hitherto lifeless earth springs into being; gross and flowering plants appear on every hand, grown to the height of a horse's head" (Sears, 1895, p. 262). He also recounted the local history. In 1531, when Spanish conquistador Francisco Pizarro began his conquest of Peru, the septennial rains proved to be his ally. He apparently landed with his troops along the northern Peruvian coast during a wet (read this now as El Niño) year. This enabled him to find food and water along the route of his conquest – food and water that in most years would not have been available. Sears wrote the following:

> it rains on the northern coast of Peru only once in seven years. All the remaining years are utterly dry. Pizarro could not have gone from the Tumbez, where he first landed after his fight with the Indians at Puna, to the valley of Tangarara and found feed for his animals, nor would he have found the little settlements mentioned as existing along the road in any other season than a wet one. Again, he would not have been driven from the Chira Valley, Tangarara, by malarial fevers, save in a wet year, as it exists in that valley only in wet years.
>
> (Sears, 1895, pp. 263–4)

Interestingly, in 1985, following the 1982–83 El Niño, several articles were written for *Disasters*, an international journal of disaster studies and practice, concerning the impact of El Niño on human health. One of the papers discussed the research finding that outbreaks of malaria (as well as of dysentery and cholera) tend to accompany El Niño events. Again, in early 1995 newspaper articles appeared on the recent research findings about links between El Niño and health effects. Sears, however, a hundred years earlier had exposed the local belief in the connection between these septennial rains (a proxy for an El Niño event) and malarial outbreaks in northern Peru. Such linkages identified in anecdotal comments are apparently being "rediscovered" today by researchers in a variety of fields. Perhaps researchers should place more value on reviewing anecdotal historical information than they presently do.

William Quinn and his colleagues strongly believed in the use of historical information of all kinds. Any information that provided clues about possible El Niño events were taken into their consideration. This enabled Quinn and his associates to construct a chronological history of El Niño events for the past several hundred years. The chronology was based on observations and historical records as found in personal diaries, ship's logs, rainfall records, and so forth. Later, they extended the chronology using more distant locations of alleged consequences of El Niño such as Nile floods, with records that go back to the seventh century.

Paleoecological evidence

There have been several attempts by researchers to reconstruct El Niño's history well back into prehistoric times. These studies, referred to as paleo studies, seek to identify indirect signs of the occurrence, frequency, and magnitude of El Niño events of past ages. Some researchers have even tried to identify more specific characteristics of those events using proxy data, which include the results of analyses of tree rings, ice cores, fossil soil deposits, marine sediment records of fish abundance, evidence of widespread fires, floods, droughts, and so forth. These types of data provide an indirect indication of the occurrence (and possibly strength) of past El Niño episodes, the local environmental impacts, and their possible teleconnections (see Diaz and Markgraf, 1992).

Clearly, such information has been quite instrumental in identifying changes over long periods of time in the characteristics of El Niño events. However, there are several assumptions that need to be validated: (a) that El Niño events of the distant prehistoric past were similar to those of today; (b) that their patterns of teleconnections, as we know them today, would have been the same under the different atmospheric and oceanic conditions occurring then; and (c) that they responded to the same forcing conditions, despite centuries-long warm and cold periods with regard to global average temperatures. (Would a human-induced global warming evoke ecological changes similar to those that would accompany a naturally occurring climate change?) There is also a problem with information generated by intervening events that could either mask the occurrence of an El Niño event (such as might occur in periods of active volcanic eruptions around the globe) or produce similar impacts in the absence of an El Niño event. In statistics, this is referred to as a problem of attribution, trying to identify causal relationships when there are many potential contributors to an environmental change.

These words of caution notwithstanding, interest in reconstructing El Niño's past history has been quite fruitful and has led to an improved scientific understanding of its causes and impacts, as well as changes in its characteristics.

Biological evidence

The consequences of the 1957–58 El Niño captured the attention of scientists at a symposium in California in the late 1950s. Symposium participants highlighted the importance of biological indicators of changes in the ocean environment, noting that "It was abundantly evident ... that the strongest and most spectacular evidence of marked change in the coastal countercurrent ... came from biological, rather than physical, observation" (CalCOFI, 1959). Initial changes in oceanic conditions may have seemingly negligible, but nevertheless detectable, impacts on some types of living marine resource, well before other more easily measured indicators become evident to researchers.

Several potential and actual biological indicators can be used to identify the possible onset of an El Niño event. One such indicator is the age structure of fish catches. During the 1972–73 event, pockets of cold water appeared close to the coastline in the midst of the general sea surface warming along the Peruvian coast. Such pockets contained large numbers of anchoveta that were easily accessed by fishermen. The fishing boats pulled anchoveta into their boats at very high catch rates and ferried their catches back to the fishmeal-processing factories along the coast, before returning later (sometimes in the same day) for additional catches. Unfortunately, few were closely watching the characteristics of the catch; adult fish were being caught in extraordinary numbers, even though this portion of the population was expected to produce future generations. By the time researchers realized what had happened, the anchoveta population was in the midst of a collapse from which it would not recover for many years.

Another biological indicator of the onset of an El Niño relates to the fat content of fish. During El Niño events, fat content is reduced as the fish draw on their fat reserves for sustenance in the presence of nutrient-poor surface waters. Linked to this is yet another biological indicator of changing environmental conditions, the Gonadic Index. Chilean researchers used the Gonadic Index in the 1982–83 event to identify a decrease in the fertility of (in this instance) the Chilean sardine. When these fish have low fat content, little fat is available for gonadic development and, consequently, for egg production, which causes a decrease in overall fertility. Therefore, increases and decreases in fat content can be used as proxy information to identify subtle ecological changes in the marine environment.

Bird populations, too, are major indicators of environmental change related to air–sea interaction in the Pacific Ocean. During El Niño events, bird populations on various islands in the tropical Pacific, for example such distant locations in the Pacific Ocean as Christmas Island and the Galapagos Islands, abandon their colonies, fail to reproduce, and suffer high mortality rates, especially among the chicks. This may be related to changes in the local fish populations, on which the birds depend for sustenance. With warm events, nutrient availability declines, fish reproduction fails, the remaining fish disperse, and they become more difficult for birds to locate. Thus, the redistribution and reduction in the amount of nutrients that reach the sunlit zone of the ocean around such islands and coastal areas, a feature that was prominent during the 1982–83 El Niño event, have been linked to the observed reproduction failure of sea birds and marine mammals on the Galapagos Islands.

> A recent investigation, utilizing satellite ocean color observations and complemented with coincident oceanographic measures, has demonstrated the tight coupling that exists between the distribution of phytoplankton populations around the Galapagos Islands and the oceanographic conditions observed during the 1982–83 El Niño.
>
> (Feldman, 1985, p. 125)

Yet another indicator of El Niño is the presence of warm-water species in regions normally occupied by cold-water species. This is due, in part, to the weakening of upwelling processes and, in part, to the appearance of large expanses of unusually warm surface water. As a result, catches of warm water species by fishermen increase during El Niño.

Observations, monitoring, and modeling

Observations of ocean temperatures, from the surface to the greater depths, are critical. Such measurements are needed on a continual basis so that researchers can calibrate against physical reality their modeling parameters, as well as their model output and their analyses. This information is also needed by specialists in remote sensing, who use the *in situ* observations and measurements for the development and refinement of their remote sensing tools.

Remote sensing

With the advent of satellites, the potential for improving society's ability to monitor, forecast and respond to the occurrence of El Niño and cold

---

Inter-Tropical Convergence Zone

Observations of the atmospheric processes that produce rainfall are among the most important contributions to the ability of tropical countries to grow enough food to feed their people. In the tropics, these processes are localized in the region scientists call the Inter-Tropical Convergence Zone or ITCZ. This is one of the most important acronyms for those interested in problems faced by tropical countries.

The ITCZ is a region that girdles the globe where the trade winds of the Northern Hemisphere meet (converge) with those of the Southern Hemisphere. The highest ocean temperatures are associated with the yearly average location of the ITCZ. The ITCZ does not occur exactly at the equator but 5–10 degrees latitude north of it.

The convergence of the trade winds, along with the high sea surface temperatures in the region, cause a rising motion of the atmosphere, which tends to produce rain-bearing cloud systems. The normal location of the ITCZ is tied to the annual progression of the seasons. As a result, its general location and its usual seasonal behavior from one year to the next is generally known. But in any given season or year, its exact location, and several of its characteristics such as the strength of its precipitation-producing processes can vary.

To a large extent, life in many tropical countries depends on the usual (read this as "expected") behavior of the ITCZ. Human activities related to agricultural production, livestock rearing, water resources management and the like can adapt to some degree of variation in the ITCZ. Sometimes they cannot, especially in the face of extreme events such as droughts and floods. El Niño and the Southern Oscillation are phenomena that generate "abnormalities" in the behavior of the ITCZ.

---

events has greatly increased. NASA researchers Lau and Busalacchi have suggested that

> the picture of a coherent, global-scale variation [*i.e., El Niño*] did not emerge until the late 1960s and early 1970s when weather satellites began to appear on the scene. In fact ... [*Bjerknes'*] ... conclusion of large-scale, coherent variations across the Pacific was largely based on composite satellite cloudiness pictures from ESSA3 and 5.
>
> (Lau and Busalacchi, 1993, p. 281)

The most recent El Niño events since the early 1980s have been comprehensively monitored at both the earth's surface and from space. From space, various parameters can be monitored over time across large expanses of the oceans. Lau and Busalacchi have reflected on the degree of success of, as well as the problems confronted by, a growing dependence on the remote sensing by satellite of the key El Niño parameters: sea surface temperatures, latent heat and moisture fluxes (this relates to rain-producing atmospheric processes and to the atmospheric cooling or warming of the ocean), atmospheric columnar water vapor (this relates to the rain potential of the atmosphere and the dynamics of the ocean and atmosphere), surface wind stress (winds that drag the upper layers of the ocean), ocean circulation, and changes in sea level (Lau and Busalacchi, 1993, p. 282). Extolling the value of remote sensing, these researchers conclude that

> Because of the close coupling of these atmospheric and oceanic variables, the simultaneous and coordinated monitoring of the entire ocean-atmosphere system by different satellite instruments is paramount.
>
> (Lau and Busalacchi, 1993, p. 291)

In other words, by using satellites, researchers can observe the results of the integration of numerous environmental processes without having to focus on and observe each process in isolation. One good example, among many, of the potential "power" of remote sensing as a monitoring tool is the ability to measure indirectly (i.e., by proxy) thunderstorm activity by monitoring outgoing longwave radiation (OLR). Briefly, the incoming energy from the sun (solar radiation) reaching the earth's surface is reemitted to space as longwave radiation. OLR can be intercepted by cumulus cloud cover. Cumulus clouds are good candidates for producing rain. Therefore, in those regions where there is such cloud formation, OLR readings are lower than those regions of the Earth's surface where clouds do not block the outgoing long-wave radiation. The shift of convective activity toward the central and eastern Pacific, a proxy indicator of the onset of an El Nino episode, can be captured by OLR measurements monitored by satellites.

Modeling activities

In the past, information about the Pacific Ocean's behavior came from *in situ* measurements from stationary and drifting buoys and from ships on their way from one location to another. And, although the satellite era has ushered in new tools and new ways to measure various aspects of the ocean and atmospheric environments, neither old nor new methods can be relied on totally to provide all the pieces to the El Niño puzzle. The former methods would take too long and are too costly to gather information over such a large expanse and across the decades and centuries in order to observe a statistically significant number of El Niño events. With regard to the latter, limits to remote sensing technology, such as an inability to make direct observations below the sea's surface, also make certain information less reliable. Because of these limitations, a growing number of researchers believe that we must rely to a great extent on computer models in order to gain insight into El Niño processes for forecasting, as well as for other El Niño-related scientific research purposes. Although such models are by definition necessarily simplifications of reality, scientists have identified mathematical expressions that represent what they consider to be key aspects of the ocean, the atmosphere, and their interactions. The models are then tested to see whether they are able to reproduce the occurrence of past El Niño events. Once satisfied that these models can reproduce the precursor stage of El Niño with some degree of reliability, modelers then turn their efforts and models toward (a) projecting the onset of future El Niño events, (b) attempting to understand how the atmosphere and the ocean interact to produce the El Niño phenomenon, or (c) identifying teleconnections in the tropical and extratropical regions.

There are different types of model, from the simple to the very complex. No single model can capture all aspects of the observed events. In general, the most widely used models are called coupled ocean–atmosphere models. These types of model are used to capture the dynamic interactions between oceanic and atmospheric processes and fall into three main classes:

- *Limited-area models.* These models focus on a relatively small geographic region within the Pacific Ocean and on a corresponding limited part of the atmosphere, rather than on either the entire globe or even the entire Pacific basin. Typically, the areal coverage of this model is the tropical part of the Pacific basin. These computer models are relatively simple in their mathematical formulation. They are easy to run on modern microcomputers, and are therefore economical. One of these limited-area models was used by oceanographers Cane and Zebiak in 1986 to successfully forecast the 1986–87 El Niño. Some researchers see it as a good candidate for making operational forecasts of El Niño. A number of investigators have used limited-area models to investigate the development and effects of internal ocean waves (i.e., Kelvin waves and Rossby waves (see pp. 54, 56)) in producing either El Niño or cold events. This type of model has been criticized because its relative simplicity and its limited domain (that is, the geographic area of coverage) exclude processes, such as the possible influence on the tropics of regions outside the tropics, that may prove to be important for understanding the El Niño phenomenon.

- *Global atmospheric general circulation models* (GCMs). One type of a global atmospheric circulation model is linked by mathematical formulas to a limited-area ocean model. Typically, the limited-area ocean model is highly detailed and capable of correctly reproducing internal ocean wave dynamics. Because the ocean part of the model has a limited geographic scope, there are often problems in capturing oceanic processes that occur outside the model's geographic domain, particularly those processes in the western Pacific Ocean.

• *Coupled general circulation models.* Another type of global atmospheric GCM is one that is linked to a global ocean GCM. These coupled models are particularly useful for generating global warming or cooling scenarios and for identifying their possible impacts on various aspects of an El Niño, such as changes in its intensity, magnitude, and duration. However, these models are very expensive to run, requiring large amounts of time on supercomputers. So, tradeoffs have to be made by scientific researchers between a model's mathematical complexity, which may better represent the real world, and the cost of running relatively simpler models for shorter periods of time.

At present, no single type of model is capable, by itself, of capturing *all* aspects of observed El Niño events. As most of these computer modeling activities are relatively recent undertakings, and as each of the models has significant limitations, their use and/or application for forecasting or for societal impacts research purposes is much debated within the scientific community. In particular, there is discussion by modelers and policymakers alike as to whether these models should be used either to make seasonal or long-lead El Niño forecasts on an operational basis, or to provide estimates of how the El Niño process might be affected by global warming. Computer models, as relatively new research tools, will likely undergo rapid improvement during the next few years. Therefore, results related to seasonal forecasts must be viewed in the context of the errors contained within the present-day models, underscoring the fact that great care should be taken by those who seek to use such model output for decisionmaking purposes. Nevertheless, modeling El Niño provides hope for the future with regard to forecasting the event with sufficient lead time to enable society to prepare for its possible consequences.

Section III
Who cares about El Niño – and why

9 International science

Preparing this chapter has been the mental equivalent to trying to walk a tightrope. I say this in the sense that each researcher contacted tended to believe that the research program in which he or she has been involved should be listed among the important contributions to El Niño research. In some sense, each is correct: broadly speaking, each field project, each modeling effort, each monitoring program has contributed to the improved understanding of oceanic and atmospheric processes as well as their interactions; those contributions cannot be minimized. Scientific investigation is a continuous process. However, to include them all would leave the reader with an alphabet soup of program acronyms and initialisms: EPOCS, GARP, GATE, GCOS, GOOS, WCP, WCRP, WCIRP, IGBP, WOCE, GEWEX, EOS, MTPE, and so forth.

I have chosen, however, to highlight only a few of these research activities that *I*, as a social scientist interested in the use by society of scientific information, considered to have been most important for improved societal awareness and understanding of El Niño.

Where have we been?

It could be argued that the first international attention that El Niño received was in the 1950s, as a result of an accident of history. The international community, although divided at the time between colonized and colonizers, capitalists and socialists, haves and have-nots, and Communists and non-Communists, came together in the 1950s to set into motion a plan to cooperate in the name of science to study the earth. This truly international scientific endeavor was known as the International Geophysical Year or IGY (Figure 9.1), although it lasted more than a year. The research findings about the atmosphere and the oceans and other scientific observations that came out of this very successful global program led to a series of international activities centered on the equatorial Pacific Ocean. A bonus to scientists was that the 1957–58 El Niño happened to occur during this period of intensive international research collaboration.

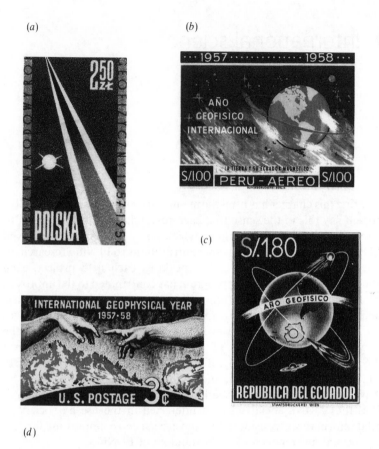

Figure 9.1. Postage stamps commemorating the International Geophysical Year.
(a) *Poland*, (b) *Peru*, (c) *Ecuador*, (d) *USA*.

In 1957, the Scientific Committee for Oceanographic Research (SCOR) identified three long-term fields of investigation with respect to the ocean environment, which were then adopted as IGY objectives: (a) the use of the deep ocean for the dumping of radioactive waste; (b) an improvement in knowledge concerning the ability of the ocean as a source of protein to feed the Earth's growing populations; and (c) the role of the oceans in climate change. With regard to their first concern, the ocean dumping of radioactive waste, scientists believed at that time that there were unexplored parts of the deep ocean that might provide suitable burial grounds for radioactive waste. In theory, this would remove the waste from any possible contamination of fish populations resulting from container erosion and the leakage of radioactive material. The second objective – exploring food production prospects from the ocean environment – was geared to the identification of regions of the world's oceans that were biologically productive and to the study of ocean currents and processes (including upwelling) that sustain relatively high levels of productivity up the food chain. In the minds of some key scientific figures involved in designing the IGY was the collapse of the California sardine fishery, an industry that was immortalized by American writer John Steinbeck in his novels *Cannery Row* and *Sweet Thursday*. The fishery declined in the 1940s, following its heyday in the previous decade, only to collapse completely by the early 1950s.

The third research objective identified by SCOR in 1957 focused on the role of the oceans in climate change. At that time, climate was not considered to be as important a politico-scientific issue as it is today. Researchers were, for the most part, unconcerned about the impact of human activities on global climate. They were interested in identifying the components of the climate system and their interaction. *New York Times* science writer Walter Sullivan highlighted the importance of the ocean to changes in the climate:

> In comparison with the atmosphere, the ocean has a tremendous capacity for absorbing, storing, and releasing heat. It has been calculated that, if all the heat of sunlight over a two-and-a-half-year period were stored up in the oceans, the temperature, in the depths, would rise only about one degree. Hence the oceans have been described as the stabilizing "flywheel" of the climate. Yet buried within them, by the same token, may be the seeds of future change, slowly moving to the surface with irresistible momentum.
> (Sullivan, 1961, p. 348)

It was fortunate for scientists and the public that the mobilization of the world's scientific community for participation in the IGY coincided with the appearance of a major El Niño. While this particular event had no visible impact on the Peruvian anchoveta fish population, it had devastating impacts on guano bird populations along the Peruvian coast. The anchoveta-consuming guano bird population was sharply reduced to about 16 million birds, down from an estimated 30 million. The information on El Niño collected during the 1957–58 event was very instrumental in the development of insights by Jacob Bjerknes about air–sea interaction in the Pacific Ocean and in the linking together of changes in sea surface temperatures and trans-basin sea level pressure.

The scientific interest and discoveries of the 1950s continued into the 1960s, sparking several research efforts related to the Pacific Ocean, the atmosphere, and their interactions. One of the most important breakthroughs came as a result of assessing research output from IGY-related studies of anomalous oceanic and atmospheric behavior in the equatorial Pacific Ocean. From these data, Bjerknes was able to make the following observation:

> Weakness and temporary elimination of the equatorial easterly [*west-ward-flowing*] winds over the eastern and central Pacific in late 1957 and early 1958 brought about a brief cessation of equatorial upwelling, which in turn caused the occurrence of above-normal surface water temperatures in the tropical Pacific from the American coast westward to the date line.
>
> (Bjerknes, 1966, p. 820)

Connecting atmospheric processes to oceanic ones enabled Bjerknes and scientists after him to better understand the El Niño process.

In the early 1970s, Dr Walter Orr Roberts, founder and former director of the National Center for Atmospheric Research (NCAR), was involved in the creation of a new international, non-governmental organization, the International Federation of Institutes for Advanced Study (IFIAS), based in Stockholm, Sweden. The Federation, which included more than 20 research centers from around the globe, set out to foster multidisciplinary approaches to science-related problems that plagued both rich and poor nations. Its member institutes were most concerned about bringing attention to the human aspects of scientific research.

One of the Federation's first activities was to create a climate program consisting of three research activities: (a) A review of the climate of 1972 and its impacts on human activities worldwide; (b) an assessment of the possible impacts of the occurrence of three poor rainfall years in a row on worldwide agricultural production; and (c) the value of a long-range climate forecast.

There was a sound reason behind the call for research on each of these three topics: 1972 had been a year of numerous climate anomalies. Many human activities in countries around the globe had been disrupted in major ways by the climate anomalies noted earlier: droughts plagued the USSR, West Africa, northeast Africa, southern Africa, India, Australia, Central America, Northeast Brazil, and so forth. These worldwide anomalies sparked one of the three IFIAS climate studies as a result of a belief among some scientists that perhaps the Earth was once again moving back into an ice age. In response to highly unexpected climate conditions in the early 1970s, several books appeared in bookstores to explain the bizarre weather, with provocative titles such as *The Cooling, Fire and Ice*, and *Weather Conspiracy*.

These adverse anomalies also had a major impact on food production as well as on global food reserves. Used to compensate for agricultural losses in the fields, reserve food stocks fell to dangerously low levels. Observers noted that the amount of *global* food reserves had declined to such an extent as to leave only a few weeks of grain reserves available if drought-related disasters around the globe were to continue. Fear of widespread food shortages was fueled by actual severe food shortages and famines in various locations, especially on the African continent and, within Africa, particularly the West African Sahel and Ethiopia. The apparent vulnerability of the global food supply generated interest in a second IFIAS climate study, to investigate how the international community of nations might handle three successive years of drought of the order of magnitude of those that had occurred in 1972.

In the early 1970s, the West African Sahel continued to be plagued by drought (1968–1973). Famine was widespread throughout the region. In drought situations of varying intensities, the notion that "if only we knew drought was coming, we could have prevented its worst impacts" is often heard. Such a sentiment translates into the belief that a better long-lead forecast of a climate anomaly, available months or even a year in advance, could be used to mitigate, if not avoid, the adverse impacts of a meteorological drought. This notion was the catalyst for the third IFIAS climate study to identify the value to societies of long-lead forecasts.

The objective of the third study, however, was to focus on the societal needs for, and constraints on, the use of meteorological information about certain kinds of extreme meteorological events. As drought was seen by many as *the* main constraint to food production at that time, it was logical to focus on the value of a long-range forecast of drought in the West African Sahel.

The result of that study – that the absence of meteorological forecasts was not the main constraint on enhancing food production or in averting food shortages in Africa, but societal factors dominated – sparked a call by some atmospheric scientists to show the value of such a long-range forecast in an industrialized country. In response to this call, a companion study on the value of a long-range forecast was undertaken for the spring-wheat region of Canada. The results of this study also suggested that an improved long-range forecast for the region would be of limited value, in the absence of the necessary changes in the ways that society prepared for drought-related agricultural production shortages. Wheat farmers in the Canadian Prairie Provinces were already using weather and climate information to make farming decisions in order to hedge their bets against the possibility of climate-related problems.

Given that Peru's coastal waters are among the most productive in the world, and given that occasional El Niño events interrupted high levels of fish catches along that coast, it seemed useful to assess the value to Peru of a long-range forecast of El Niño. In a study to assess the value of a long-range forecast of the major 1972–73 El Niño event, experts were asked to suggest how they might have reacted to a forecast of the 1972–73 El Niño a year in advance of its onset. For this study, many researchers and policymakers were asked for their views on what they perceived to be the value, at least in theory, of such a forecast. Those surveyed about El Niño included scientists, representatives of the fishing industry, political representatives, and so forth. One of the most interesting findings of this study was that El Niño forecasts could clearly be of benefit to Peruvian decisionmakers. However, existing political, economic, and social constraints (at that time) inhibited the prospects for a full realization of that benefit (Glantz, 1981).

The International Decade of Ocean Exploration

In the 1960s, coastal upwelling regions were recognized by the International Biological Program Biome Committee as one of the world's major types of ecosystem. They are considered to be important in the sense that they are among the most biologically productive regions in the global ocean. A large portion of commercial fish landings are taken from these highly productive regions each year. The Coastal Upwelling Ecosystems Analysis Program, CUEA for short, was developed for the International Decade for Ocean Exploration (IDOE) under the auspices of the United States National Science Foundation. At a cost of more than $US20 million, it was designed to improve understanding of the chemical, physical, and biological aspects of coastal upwelling regions "well enough to predict its response far enough in advance to be useful to mankind" (IDOE, n.d.).

More specifically, CUEA's goal was the development of an ecological systems model through the development and coupling of a physical oceanography systems model with a biological systems model. The coupled model would then serve to improve the ability to forecast upwelling and, therefore, to produce reliable forecasts of the locations of the most productive fisheries.

In the period between 1972 and 1977, CUEA researchers undertook six field activities in four geographic areas off the coasts of Oregon, Baja California (Mexico), northwest Africa, and Peru. The program involved scientists from 15 institutions around the world and from disciplines such as meteorology, oceanography, and biology. Scientists monitored such variables as ocean currents, waves, sea surface temperatures, winds, chemical concentrations in the sea, and the uptake of nutrients by plankton. About 200 reports and papers were published throughout the 1970s on research findings and observations resulting from CUEA's activities, with a summary of the scientific findings published in 1981 (Richards, 1981).

According to Feenan Jennings, a key science administrator for the IDOE at the National Science Foundation, Vice President Spiro Agnew had intervened in 1969 during the planning phase for setting up the IDOE to remove plans to apply CUEA's research findings to societal needs (Jennings, 1981). Agnew specifically opposed the stated objective of improving the management of fisheries. On this issue, Jennings noted:

> When the IDOE was established in 1969, the goals set forth by the Vice President did not include any direct mention of living resources. As it

turned out, this was no accidental omission. The Office of Management and Budget had insisted that programs associated with fisheries or food from the sea be left out of the IDOE charter. They were negotiating with the US Bureau of Commercial Fisheries about its programs and did not want any new living resources programs started in the federal government until the issues were resolved.

(Jennings, 1981, p. 13)

Two IDOE scientific advisory groups, however, recommended that living marine resources be included. By 1971, the Office of Management and Budget had dropped its opposition and agreed with them. As Richard Barber, a biological oceanographer and key participant in CUEA, once noted,

> The long-range goal of the CUEA program includes protecting the long-term economic productivity of the ecosystem by providing a basis for relating management of the fisheries to the natural climatic and oceanographic variability and other environmental perturbations. In this manner maximum social benefits can be obtained from the natural resources in upwelling regions, both now and in the future.
>
> (Barber, 1977, p. 1)

Although the CUEA project was multidisciplinary with respect to the natural sciences (i.e., biology, chemistry, physical oceanography, meteorology) and although statements were made such as "Amongst the oceanic phenomena affecting man, coastal upwelling has the most direct and largest impact on the social fabric" (IDOE, n.d.), there was apparently no plan to investigate the political, economic, and environmental implications of the availability of a reliable upwelling forecast.

CUEA was not set up to investigate El Niño events, although the Peruvian upwelling system was targeted for CUEA-related research. CUEA, however, was by chance carried out during an El Niño episode, during which there was a major disruption of the coastal upwelling processes along the coast of western South America. While brief statements appeared in project reports alluding to the economic and social importance of an upwelling forecast, as well as in program descriptions and memos pertaining to the CUEA project, such statements were based on either a view that such forecasts could have only positive benefits for humanity or that there was a need for justification of the science program in societal terms. An example of this optimistic reasoning was captured in an article (*Mosaic*, 1974) summarizing current (as of that time) scientific research efforts on upwelling and, by implication, El Niño forecasts, suggesting that

1. By gaining a deeper understanding of the physical and biological dynamics of upwelling, scientists, fishermen, and national legislators will have more potential control over the supply of fish stocks.

2. The major practical benefit of an upwelling prediction system for locations throughout the world is that fishermen will be able to obtain an above-average harvest with less time wasted searching for fishing locations.

(*Mosaic*, 1974, p. 31)

However, the fact that minimal attention was paid to the social application of CUEA's scientific findings was eventually acknowledged in an IDOE report.

When IDOE was established, its objective of achieving more comprehensive knowledge of ocean characteristics and more profound understanding of oceanic processes was linked to the more effective utilization of the ocean and its resources. ... In practice, the IDOE investigations have leaned toward fundamental research.

(FAO, 1974, p. 5)

While calling for continued emphasis on fundamental research, the report asked scientific researchers to "*keep in mind* the areas of eventual application of research findings" [italics added].

During a wrap-up meeting on the contributions of CUEA in 1980, a participant from the social sciences seemed hard-pressed to identify direct benefits of CUEA's scientific research endeavors to society. He did suggest that society would ultimately benefit from its basic research findings.

Today, CUEA may be only a faint memory in the minds of its participants. However, it was clearly an important research and field program in terms of enhancing our understanding of the role of coastal upwelling ecosystems in affecting regional climates in general and, more specifically, in year-to-year global climate variability with regard to changes in the equatorial Pacific upwelling regions, both along the equator and the coast.

ERFEN

In 1974, following the adverse impacts on the region's ecosystems and economies, four South American countries – Chile, Peru, Ecuador, and Colombia – created ERFEN. ERFEN is the acronym for the Estudio Regional del Fenomeno de El Niño (in English, the Regional Study of the El Niño Phenomenon).

ERFEN conducts cooperative multidisciplinary research involving the physical, biological, and social sciences. Seventeen institutions in the region participate in the research. Its overriding objective has been to undertake El Niño-related research in order to predict changes in the interactions between air and sea for the purpose of implementing policies for fisheries, agriculture, industry, and public safety.

During its more than 20 years of activity, ERFEN has been instrumental in generating awareness of the potential value of understanding and forecasting El Niño. It has published a monthly climate awareness report that is widely distributed in the region by the Permanent South Pacific Commission (CPPS, in Spanish).

The First World Climate Conference

During the 1970s, the United Nations organized several world conferences on important development-related issues. They began in 1974 with the World Food Conference (in Rome, Italy) and the World Population Conference (in Bucharest, Romania). These were so successful that they were followed by a Conference on Human Habitat. This was followed by conferences in 1977 on water (Argentina) and desertification (Nairobi, Kenya), and conferences in 1979 on technology (Vienna, Austria) and on climate (Geneva, Switzerland). National representatives to the 1979 World Climate Conference called for the creation of the World Climate Program, to be based in the World Meteorological Organization in Geneva. The climate program was divided into four parts: the World Climate Research Program, the World Climate Data Program, the World Climate Applications Program, and the World Climate Impacts Program. The first three were under the auspices of the World Meteorological Organization, and the last one was given to the charge of the United Nations Environment Programme (UNEP). The impacts program is the only part of the World Climate Program that is directly concerned with the interactions between climate and society.

In 1985, UNEP established an international, multidisciplinary working group on the social and economic consequences associated with El Niño events. This working group served as an advisory group and identified important El Niño-related issues. The Working Group organized several El Niño-related workshops and produced several publications on the physical, biological, and social aspects of El Niño.

Where are we now?

The Tropical Ocean–Global Atmosphere (TOGA) Program

The TOGA program was an ambitious observations and modeling program focused on air–sea interactions in the equatorial Pacific and their influence on global climate variability. TOGA was developed in the early 1980s under the auspices of the United Nations' World Climate Research Programme and ran for ten years, from 1985 to 1994. According to the TOGA implementation plan,

The scientific community has been aware of the existence of anomalous oceanic and atmospheric circulation patterns that can develop on time scales of several months to several years and has recognized that a significant part of these variations can be explained by the dynamics of the coupled [*i.e., air and sea*] system constituted by the tropical ocean and the global atmosphere.

(TOGA, 1987, p. 7–1)

The most prominent manifestations of air–sea interaction in the Pacific region are El Niño and the Southern Oscillation. The general character of the two-way interactions between the atmosphere and the ocean are displayed schematically in Figure 9.2.

TOGA's observational tools included satellites, ships, moored and drifting buoys, and tide gauges, and were designed to create reliable records

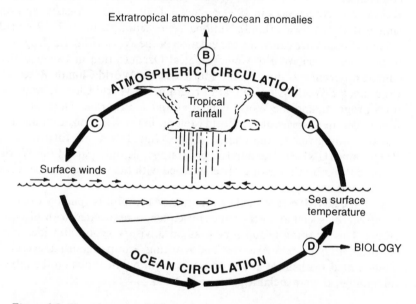

Figure 9.2. Key elements of concern to TOGA researchers. The distribution of rainfall in the tropics and the large-scale atmospheric circulation are highly sensitive to the distribution of sea surface temperature (A). The pattern of tropical rainfall, in turn, influences the position and intensity of jetstreams and stormtracks in extratropical latitudes (B), and the surface wind field over the tropical oceans (C). These surface winds, in turn, drive the current systems in the upper layers of the equatorial oceans through the stress at the air–sea interface. The currents determine the distribution of upwelling (D), which is of vital importance to marine biology and fisheries and plays a major role in determining the distribution of sea surface temperature, which feeds back on the atmospheric circulation. (From NRC, 1990.)

over time for several variables such as sea surface temperature, wind, upper ocean temperature structure, ocean currents, and sea level. This information was necessary and important for future attempts to predict short-term climate variability within as well as outside the tropics. It was also of great value to modeling efforts geared to developing long-range prediction capabilities. (An example of the array of measuring devices is shown in Figure 9.3. The Tropical Atmosphere Ocean Array of measuring devices is a valuable legacy of the recent TOGA program. It makes possible real-time forecasts of, for example, changes in wind direction in the Niño3 area available to the El Niño research and modeling communities.) In sum, TOGA aimed "to explore the predictability of the tropical ocean–atmosphere system and the impact on the global atmospheric climate on time scales of months to years." The identification of teleconnections between the 1982–83 El Niño and weather anomalies in North America also strengthened the resolve of United States scientists during TOGA to better understand and model air–sea interactions in the Pacific.

As happened during the IGY and during CUEA, a major El Niño event took place during the early planning phase for TOGA. This El Niño event gave strong impetus and a sense of urgency to the need to carry out TOGA. Among TOGA's achievements were a better understanding of the El Niño life cycle, the capability to monitor climate in real time and global pictures of El Niño on a routine basis.

Within the TOGA program, the Coupled Ocean-Atmosphere Response Experiment (COARE) was an unprecedented experiment designed to focus on the western Pacific Ocean region known as the warm pool. The warmpool, home of the world's warmest open ocean sea surface temperatures, has been referred to as a "heat engine" for atmospheric processes. TOGA-COARE was a data collection effort to determine how warm water spreads across the Pacific basin. It involved more than a thousand scientists from many countries, and the use of seven satellites, 14 ships, numerous automated ocean buoys, and enhancements to upper-air weather stations. The field activities took place in the western Pacific from October 1992 to February 1993. The data gathered were used to help climate and ocean modelers in their efforts to better model air–sea interactions and perhaps to better identify their consequences for the purpose of improving long-range (e.g., long-lead) weather forecasts. The goal of this field program was to observe the air–sea interactions on a variety of time scales, from hours to weeks to months, in an attempt to better understand the natural variability of those interactions.

Climate variability (CLIVAR)

CLIVAR is a newly established 15 year scientific research program designed to study climate variability and predictability. It is a direct

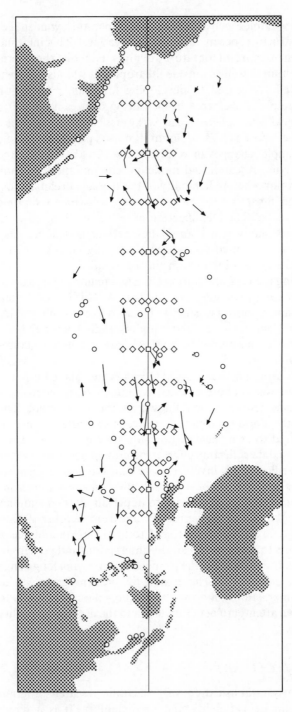

Figure 9.3. Shown here are some of the components of the ocean observing system that are being deployed in support of El Niño prediction. The open circles denote automatic tide gauge stations. The squares and diamonds indicate the locations of moored buoys, which monitor surface wind and other climate elements, as well as water temperatures at several levels below the ocean surface. They operate continuously for months at a time, without human intervention. The arrows depict the tracks of drifting buoys, which measure water temperature and reveal the motion of the surface water. Many of these observations are sent directly to weather prediction centers around the world via satellite. (Graphic and caption courtesy of Interwork, Inc., Del Mar, California, USA.)

follow-up to the TOGA program, which ended officially in December 1994. The concept of CLIVAR, as stated,

> arose from the recognition that observed climate variations result from natural variability superimposed on long-term trends that may be induced by anthropogenic modification of the global environment and other external forcing factors.
>
> (WMO, 1995)

To address these concerns, including the possible impacts of a human-induced climate change, CLIVAR was organized around three scientific objectives: (a) to study seasonal-to-interannual climate variability and predictability; (b) to examine the role of the ocean in decadal-to-centennial climate variability; and (c) to model and detect anthropogenic climate change.

The scientific objectives of CLIVAR are as follows:

- to describe and understand the physical processes responsible for climate variability and predictability on seasonal, interannual, decadal, and centennial time scales, through the collection and analysis of observations and the development and application of models of the climate system;
- to extend the record of climate variability of the time scales of interest through the assembly of quality-controlled paleoclimatic and instrumental data sets;
- to extend the range and accuracy of seasonal-to-interannual climate prediction through the development of global coupled models.

> (WCRP, 1995, p. 3)

Thus, as a follow-up to TOGA findings, El Niño research is one of the three key elements of CLIVAR.

Climate change and El Niño

There is a proverbial dark cloud hovering on the horizon for the El Niño forecast community. That dark cloud represents the uncertainty surrounding the possible changes in El Niño patterns and characteristics that might accompany global climate change. Interest has been growing since the late 1970s in the potential implications of increased emissions of greenhouse gases for atmospheric processes. That interest has focused, for the most part, on the effects on the global climate of greenhouse gas emissions. These gases include carbon dioxide and methane produced from growing wet rice, nitrous oxides linked to fertilizers, and chlorofluorocarbons (CFCs) used in aerosol sprays, foam-blowing agents, and refrigerants.

Numerous climate scenarios have been derived, using large computer models to generate possible regional consequences of projected global warming of the atmosphere somewhere between 1.0 and 4.5°C that would probably result from a doubling of atmospheric carbon dioxide and other greenhouse gases over the level that existed at the outset of the Industrial Revolution in the mid-1800s (Houghton *et al.*, 1990).

As most climate model results have described the possible changes of an average climate, many of the climate impact studies have used model results as a point of departure to address possible average environmental and ecological changes due to future increases of greenhouse gases. Climate impacts researchers have used the model-based scenarios in their attempts to identify possible climate-related environmental changes with which future societies might have to cope (Parry *et al.*, 1988).

Impacts researchers have pursued other lines of reasoning as well in order to gain a glimpse of possible societal responses to climate-related environmental changes. Few, however, have examined possible changes in extreme weather events, mainly because the model studies usually emphasize changes in average climate conditions and not changes in climate variability. Yet, El Niño events are associated with extreme climate-related anomalies in various regions around the globe.

Possible changes in the key characteristics of El Niño events that might result from global warming were discussed only briefly in the first scientific assessment of the United Nations-sponsored Intergovernmental Panel on Climate Change (IPCC) (Houghton *et al.*, 1990). More recently, models used to study a carbon dioxide-induced climate change include more realistic oceanic conditions that can change over time. However, as of now, only one of these models has been able to generate, still rather poorly, El Niño-like conditions. Changes in extreme climate events associated with El Niño should receive more attention.

It is now accepted that El Niño affects climate on a global scale. In the tropics, the strongest effects often involve widely scattered regions, stretching from the eastern coast of the African continent to the Indian subcontinent to Southeast Asia, and across the Pacific Ocean to various locations on the South American continent. As noted earlier, there is evidence that El Niño has affected climate patterns in distant parts of the Northern Hemisphere, as far away from the equatorial Pacific as North America, Russia, and the Ukraine (e.g., Flohn and Fleer, 1975).

Studies have been undertaken of earlier warm periods, such as that which occurred in the Medieval Optimum period (around AD 1200), or historical assessments of the warm decades of the 1920s and 1930s, to identify changes in El Niño behavior related to changes in climate. Fisheries expert Andrew Bakun (1990, p. 198) speculated that global warming could intensify the winds that blow along the shore of the west coast of continents, i.e., the upwelling zones. How an increase in the intensity of coastal upwelling processes in the eastern equatorial Pacific might affect fish populations was not clear.

At a UNEP-sponsored workshop on climate change and El Niño, held in Bangkok, Thailand, in 1991, Australian meteorologist Neville Nicholls posed questions to other participants about what one might be able to say about the possible changes in El Niño resulting from climate change, given the state of our knowledge at that time (Glantz, 1991). The consensus of the workshop participants is presented below in response to the following questions that are constantly being asked.

Q: *Will El Niño continue to occur?*
A: Likely. Proxy climate evidence indicates that El Niño has continued to occur in a variety of different climate regimes, both warmer and colder. There is little evidence that El Niño has been "stuck" in one extreme.
Q: *Will El Niño be more intense?*
A: Possibly. Global coupled model results suggest an intensification of El Niño climate effects in the tropics, with anomalously wet areas becoming wetter, and dry areas drier.
Q: *Will El Niño change in frequency?*
A: Possibly. [According to Trenberth and Hoar (1996), "the tendency for more frequent El Niño events and fewer La Niña [cold] events since the late 1970s has been linked to decadal changes in climate through the Pacific basin. ... Both the recent trend for more ENSO events since 1976 and the prolonged 1990–1995 event are unexpected given the previous record, with a probability of occurrence about once in 2,000 years. This opens up the possibility that the ENSO changes may be partly caused by the observed increases in greenhouse gases."]
Q: *Will El Niño change in duration?*
A: Less likely. Present-day events exhibit a tendency toward two years in length linked to the annual cycle of the natural flow of the seasons. However, single and multi-year events have occurred in the past, and there is no evidence to indicate a change in this characteristic. This also implies that the mechanisms involved with transition from a warm to cold phase (i.e., El Niño to La Niña), and vice versa, would remain similar.
Q: *Will there be a change in tropical teleconnections?*
A: Unclear. A global coupled model suggests that teleconnection patterns similar to those of today would occur in the tropics with an increase of carbon dioxide in the atmosphere.
Q: *Will there be a systematic change in extratropical teleconnections?*
A: Probably. Prior modeling work on extratropical teleconnections and changed basic climate states suggests that a large mean climate change (e.g., increases in temperature) associated with an increase of carbon dioxide and trace gases would result in altered extratropical teleconnections. Results from a global coupled model with increased carbon dioxide are consistent with this earlier work. On the basis of analyses of records of the past 100 years or so, these teleconnections appear to have held.
Q: *If changes do occur, will there be changes in both warm and cold events?*

A: Probably. Changes in upper ocean thermal structure are likely to affect both warm and cold extremes of the oscillation. Associated changes of sea surface temperature and convection have implications for effects on the monsoons and typhoon activity.

At another workshop held a few years later in Australia, the same questions were raised. Responses similar to those at the Bangkok workshop were given by a different collection of participants.

Discussion among scientists about the implications of climate change on various aspects of El Niño in its broadest meaning, as a Pacific basin-wide phenomenon, is highly speculative at this time. El Niño forecasters and researchers alike have felt somewhat baffled by the recent occurrences (1991–95) in the equatorial Pacific. The pattern of changes of sea surface temperature and sea level pressure in that period were unusual and, while unexpected, were clearly not unprecedented. Therefore, it is important to reaffirm the warning issued in the report of the Australian workshop on *Climate Change and El Niño-Southern Oscillation*:

> The participants at the meeting agreed that our understanding of the El Niño–Southern Oscillation is still far from complete, and the modelling of the phenomenon is still rather rudimentary. Much work is still required to refine and confirm the possible effects of climate change on the El Niño–Southern Oscillation, before they can be used in assessing societal or ecosystem impacts.

(Nicholls, 1993*a*, p. 9)

Since there is still active debate about the mechanisms that produce ENSO, it becomes problematic to model future behavior of ENSO under conditions of global warming. Research has shown that there is likely to be a suite of ENSO mechanisms, and one or more in combination may work to produce what we observe and call ENSO. The failure of the forecast models in the early 1990s seemed to confirm this notion. If all the mechanisms are not accounted for in the models, some ENSO forecasts will fail because of the absence of certain processes. Anything we say about future changes of ENSO must, at the very least, be tempered by our lack of ability not only to simulate present-day ENSO phenomena with models, but also to understand fully the observed behavior of ENSO events.

Gerald Meehl, US National Center for Atmospheric Research

Where are we going?

In a period when attention has shifted to concern about human-induced global environmental changes such as global warming, ozone depletion, desertification and tropical deforestation, there has been a tendency to downplay the importance to societies of naturally occurring short-term changes. Perhaps one reason for this is that the latter are episodic events, whereas the former types are continuing changes that are constantly present. Another reason might be the feeling that there is little to be done about the naturally occurring events, such as El Niño, but that the human-induced changes can be affected by decisionmaking processes. Yet, El Niño is a naturally occurring environmental change that has global implications.

El Niño events affects the lives of hundreds, if not millions to a billion people, either directly or indirectly, either positively or negatively, and mostly directly in the tropical countries that girdle the globe. Any information in advance about the possible onset or about the actual beginning of an event can be used by those in power to mitigate its possible adverse impacts and capitalize on its positive effects.

As far as anyone can tell, El Niño events are here to stay. Societies must learn more about them and learn how to use El Niño-related information for their betterment. The same advice applies to cold events. A society that is forewarned about El Niño will be forearmed.

The scientific research community dealing with El Niño has for the most part focused on forecasting the onset of the event. While this is very important and useful to societies potentially affected by it, it is only part of the picture. If all El Niño-related research were to be halted today, there would still be considerable value in using existing information about El Niño in decisionmaking processes in both tropical and extra-tropical countries. National examples of the benefits of improved awareness of El Niño are increasing: Ethiopia, South Africa, Canada, Brazil, Vietnam, and Cuba.

The El Niño research community is only now beginning to attract social scientists into their camp, and social scientists have only recently become interested in identifying ways to better understand and apply that understanding to the needs of decisionmakers in various sectors of society, individuals as well as national policymakers. Clearly, El Niño research is one of the bright spots on the scientific horizon, as we enter the twenty-first century.

The physical and biological research communities are poised to continue their efforts to unlock the remaining mysteries of air–sea interaction in the Pacific Ocean and their teleconnections to weather and climate anomalies around the globe. The social sciences are eager to get involved and are beginning to do so. This relatively neglected recurring natural global environmental change is beginning to get the respect it deserves.

We do not know whether the global climate will change in future decades, nor do we know when or how that change might manifest itself. Clearly, we do not yet have a clue as to how a global warming might affect the characteristics of an El Niño. If, however, governments want to gain an idea of how well-prepared societies might be with regard to coping with the consequences of climate change, they might try to gain a better view of how, as well as how well, societies today cope with climate variability. We have studies about how different groups and nations have coped with extreme weather-related events such as floods, droughts, frosts. These are often random events. However, El Niño is a recurring event, an experiment of sorts in the Pacific, which provides us every so often with information about how well or how badly societies have coped with extreme meteorological events.

El Niño research efforts present society with a "win–win" situation. Any improvement in our understanding of the event, its teleconnections and its societal and environmental impacts can be of value to society. The use of that information can also benefit society. There is nothing to lose in supporting El Niño-related physical, biological, and social science research. In addition to being an exciting field of research, it has great potential benefits. It is important that the public knows about El Niño and about the physical, biological, and social scientific research efforts proposed to improve our understanding of it.

Climate prediction institute

Considerable research effort and funding in the past ten years has been directed toward addressing climate change issues such as causes, consequences, and possible policy options. This effort has contributed to the downplaying of the importance of research on climate variability from year to year. These time scales directly affect current generations of people. It was recognized that to reap benefit from climate-related research, a more balanced research effort must be put in place, one that provides adequate resources for the investigations of both climate variability and climate change. As a key part of this balance, research on the societal aspects of climate variability and change must also be integrated into the research plan.

To improve the ability to forecast interannual variability and to realize the benefits therefrom, the United States government proposed the creation of an international research institute (Figure 9.4) for climate prediction at the Earth Summit in Rio de Janeiro, Brazil, in June 1992. Several governments have since supported a proposal to establish "an end-to-end (i.e., research to application) multinational, seasonal-to-interannual, climate prediction program (SCPP) ... based on the evolution of existing efforts to observe, understand, predict, and assess oceanic and atmospheric processes" (NOAA/OGP, 1994, p. 2). Such a program is intended to assemble participants from around the world to achieve a task no single country could accomplish on its own: the creation of a global climate forecast system with the capacity to employ regional analyses to refine forecasts and apply them for the benefit of human societies.

(NOAA/OGP, 1994, p. viii)

In November 1995, following many preparatory workshops around the globe to stimulate interest in such an institute, an International Forum on Forecasting El Niño was convened in Washington, DC, by NOAA's Office of Global Programs. High-level representatives of Ministries of Science and Technology from around the world, and their advisors, were invited to discuss ENSO and related climate variability, the effects of climate variability on society, and advances in year-to-year forecasts of ENSO-related climate variability. The forum was also expected to produce broad-based international support for the creation of a multinational mechanism to launch a worldwide network with "end-to-end" capabilities in climate forecasting and applications. The forum was, to date, the largest multinational, multidisciplinary, action-oriented gathering of people whose interests were focused on the El Niño phenomenon.

A crucial step forward in this regard was the announcement by NOAA in

Figure 9.4. Logo for forum to create climate prediction institute.

March 1996 that the Lamont–Doherty Earth Observatory of Columbia University and the Scripps Institution of Oceanography were selected to co-host the international research institute for short-term climate prediction. Within such an institute, the talents of physical and social scientists would be drawn together to support research on the coupled climate system aimed toward improved predictive skill and the regular production and distribution of experimental forecasts of climate variability on time scales from a season to a few years in advance. The goals and objectives of the institute will include the development of coupled models of the ocean–atmosphere–land climate system to be used as the basis for improved climate predictions, the routine generation of experimental climate forecasts, collection of observational data needed to improve climate forecasts, and the tailoring and application of climate forecasts for the explicit social and economic benefit of both developing and developed nations.

In the planning documents for the research institute there are numerous references to the benefits to society that would probably accrue from the implementation of an international research center for climate prediction on seasonal-to-interannual time scales. Achieving those benefits is neither a simple nor an easy task. Platitudes, hopes, and sympathetic feelings will not, by themselves, lead to those benefits. We have seen that much scientific information remains untapped for its potential contribution to wellbeing. It is a challenge that requires systematic research on the actual use of climate predictions. It will require explicit and active attention from dedicated social scientists and others interested in research applications to identify potential uses as well as users of seasonal and yearly forecasts. They can serve as the bridge builders and translators operating between the scientific community and society and, therefore, must be given an active and prominent role in the establishment of such forecast centers.

10 Why care about El Niño?

It is quite clear that several regions around the globe have been identified as areas affected by El Niño, in its broadest sense. Governments and individuals in those areas have good evidence about the linkages between El Niño and regional climatic anomalies. They can use El Niño information to improve the way they might cope with the consequences of their regional El Niño-related climate anomalies. For example, in a letter to the British medical journal the *Lancet*, Dutch representatives of Médecins sans Frontières (Doctors without Borders), Bouma *et al.*, suggested that El Niño has been the driving force behind periodic malaria epidemics. They provided the map shown in Figure 10.1 to support their contention. They went on to suggest that

> the advances made in the past decade in meteorological forecasting of the phases of the Southern Oscillation [warm and cold events] may help to predict areas at risk of malaria epidemics. ... This offers possibilities for developing early warning systems that can facilitate epidemic preparedness.
>
> (Bouma *et al.*, 1994, p. 1140)

The public, too, has shown increasing interest in El Niño. People know, when they hear the words El Niño, that something is happening somewhere in the Pacific Ocean. However, they may not know exactly what that something is. They may also believe that it has an effect on their weather, but they may not know with any degree of exactness what that effect is. For example, in the media one can find such misstatements as the "Hurricane El Niño," or such questionable comments as "the 1993 American Midwest floods were caused by El Niño."

To scientists, such comments suggest a widespread ignorance about El Niño. However, to those concerned about putting into practice the potential uses of El Niño information, such comments represent a significant starting point for educating members of the public in more detail about how changes in sea surface temperature and sea level pressure across the equatorial Pacific might affect their well-being. The potential importance of

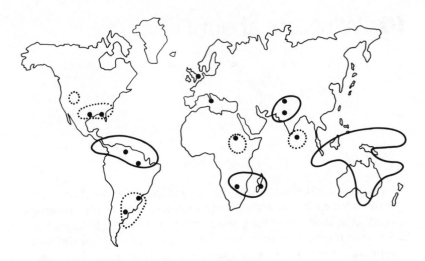

Figure 10.1. ENSO-related precipitation anomalies and periodic malaria epidemics. ...., wet regions; _____, dry regions; •, malaria, past and present. (From Bouma et al., 1994. Lancet, 343, 1140, © by The Lancet Ltd, 1994.)

El Niño has already established beachheads of awareness, so to speak, in the public mind. The best approach may be to expand from those beachheads to other individuals, agencies, and sectors of society.

While there are several regions where El Niño teleconnections appear to have an acceptable degree of reliability, there are many parts of the globe where connections to El Niño events have yet to be established reliably. In still other locations regional climate anomalies associated with El Niño are known but appear to have only minor societal or environmental impacts. And within countries that are affected by El Niño there are local areas that remain untouched by climate anomalies spawned by this particular air–sea interaction. Why, then, should people in these seemingly unaffected regions care at all about changes in sea surface temperatures and sea level pressure across the equatorial Pacific Ocean?

Consider Kenya, for example. The actual linkages of Kenya's climate anomalies to El Niño events are not very clear. Therefore, there is a tendency for Kenyan policymakers not to care much about the warming of sea surface temperatures halfway around the globe. Kenya, however, grows and exports coffee. Many international competitors of Kenyan coffee growers, such as those in Brazil, Ethiopia, or Indonesia, are more clearly affected by El Niño events that can alter their ability to meet the demands for their coffee in the international marketplace. As another example, consider palm oil. Palm oil production in the Philippines declines during El Niño events, which tend to spawn droughts in the region. As a result, commodity brokers who wish to purchase palm oil at low prices must find other sources of palm oil in, for example, West or Central Africa. If they wait until the full effects of an El Niño are felt, the price of palm oil will likely have increased due to reduced supplies from the Philippines. In fact, even a hint of the onset of an El Niño could be of value to these decisionmakers.

As another example, commodity brokers who deal in the marketing of squid issue futures contracts for the purchase of squid captured off the California coast. These contracts promise to deliver a set quantity of squid at a fixed price agreed upon many months in advance of its proposed delivery. During El Niño events, however, it appears that squid catches drastically decline off the California coast. In the absence of a forecast of an impending El Niño, commodity brokers would probably have to go to the international marketplace in order to buy squid at much higher prices than the price they would receive for it, as a result of legal contracts made months earlier. Warned of the possible onset of El Niño, however, astute commodity brokers would have the option to make additional contracts for squid purchases in other parts of the world at relatively low prices in order to meet their contract obligations. They would, then, be able to avoid being forced to purchase squid at very high prices because of the El Niño-induced scarcity of that resource in California's waters. Similar stories exist for other commodities as well.

These examples show that it is not necessary for an El Niño event to directly affect a country's economic activities in order for its decisionmakers to be concerned about gathering more information about the phenomenon and its potential regional consequences. Even the wise use of existing El Niño information, including, but not limited to, El Niño forecasts, can provide decisionmakers with a considerable amount of information with which to hedge their decisions. As Colin Ramage suggested, "Whether or not it is possible to forecast the onset of El Niño, one might still hope to predict its effects outside the tropics once it has begun" (Ramage, 1986, p. 83). This sentiment was also reflected in a letter in 1986 to the editor of *Scientific American*, signed by several key El Niño researchers:

> Because individual El Niño events evolve over a period of about two years in an established manner, once an El Niño is recognized as being under way, there is potential for adding skill to seasonal forecasting of weather and climate.
>
> (Trenberth *et al.*, 1986)

El Niño and the media

Searching for a mention of El Niño in the popular media in the USA is an interesting exercise. Going back a few decades one can hardly find a reference to an El Niño event. In those days few reporters, if any, knew about how El Niño affected climate outside the tropics. In the 1970s articles mentioning El Niño began to appear with greater frequency, as a result both of El Niño's devastating impacts on the Peruvian fisheries of the 1972–73 event and, later in the decade, of increasing scientific research activities.

The 1982–83 El Niño demonstrated to the world that El Niño could be quite devastating. Scientific research sharply accelerated in the 1980s and the media, too, showed increasing interest as well. Because of the media's interest in the phenomenon, "El Niño" was on its way to becoming a household word. Since 1983, the media have sporadically "feasted" on El Niño events and on research efforts about different aspects of the phenomenon. Stories in the media become plentiful once an event has been forecast.

Media interest in El Niño was once again heightened in the early 1990s with the appearance of what some researchers consider to have been one of the longest El Niño events in a century. This period has received considerable scientific attention because El Niño did not follow an expected pattern of decay. While it appears with hindsight that a normal El Niño occurred in 1991–92, sea surface temperatures did not drop below normal for long, and fluctuated again and again over the next three years. This anomalous behavior was partly affected by the major volcanic eruption of Mount Pinatubo in the Philippines during 1991, which caused a cooling in the tropics and elsewhere for a few years.

Regional media throughout the USA have had their interest in El Niño piqued by the possibility that their local weather might be adversely affected by air–sea interactions taking place thousands of kilometers away. That interest has been reinforced by numerous anecdotal and scientific comments on their local and regional impacts ranging from Alaska, down the west coast of North America, across northern Mexico and the Gulf region and the southeast and up the east coast of the continent into the northeast and Canada.

El Niño also provides the media with a scientific mystery. It is a mystery that could possibly be resolved by the scientific community some years into the future. The research activities of various members of the scientific community identify pieces of the climate-system puzzle. The payoff to society of increased knowledge of the El Niño process is potentially very large. That knowledge could be used to reduce the adverse impacts of climate anomalies that tend to accompany El Niño.

North Americans (Canadians as well as Americans and Mexicans) have become increasingly interested in the phenomenon and, reflecting that growing interest, the media are interested in reporting on it. The *New York Times, Boston Globe,* and *LA Times,* for example, have written feature stories about El Niño's forecasts and El Niño's impacts within the last few years. The media is an intermittent consumer of El Niño research information.

The media is also an educator of the public about El Niño. Their articles generate interest in, and respect for, the phenomenon. Therefore it is imperative that the media develops a proper understanding of El Niño: what it is, what it does, what value there is to knowing about it either in advance or once it has set in. It is also imperative that the media improve its understanding of how this information can be used by different sectors in society.

The media in other parts of the globe, where people are fearful of the possible impacts of El Niño, also give it the fullest attention. The Brazilian media, for example, is concerned when an El Niño has been forecast. Brazilians tend to be plagued by severe droughts in the poverty-stricken semiarid Northeast and by heavy floods in the south. Southern African and northeast African policymakers have also become aware of the devastation that can accompany an El Niño. The media in Southeast Asian nations, such as Vietnam, Thailand, Indonesia, the Philippines and Australia also pay close attention to El Niño because of the potential devastation an event can bring.

Droughts and El Niño

Drought is among the most dreaded of the climate-related problems faced by society. It can reduce levels of food production in rich and poor countries alike. It can exacerbate existing social and economic problems, thereby generating severe food shortages and even famine. Figure 10.2 depicts droughts that occurred around the globe in the 1982–83 period. Droughts have been implicated in desertification processes; that is, the creation of desert-like conditions where none had existed in the recent past.

Figure 10.3, an idealized version of the agricultural food production system in sub-Saharan Africa, was produced in the early 1980s by the United States Department of Agriculture (USDA) for the United States Agency for International Development to help the agency guide its food assistance policies. In the lower right hand corner is a box for weather. Weather, according to this chart, affects only crop yields. If one were to believe that, one would tend to look for ways to protect crop yields from weather variability (excessive as well as poor rains). For example, a farmer might want to change crop variety to one with a grain that might be less vulnerable to moisture stress. While the chart is more than a decade old, this narrow view of how climate affects agriculture and related human activities is still widely held.

However, what would happen to the linkages between the weather box

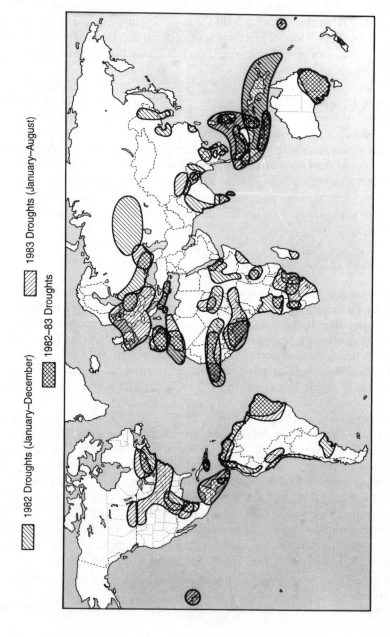

1982 Droughts (January–December)   1983 Droughts (January–August)

1982–83 Droughts

Figure 10.2. Drought map for 1982–83, highlighting the geographically widespread appearance of this naturally occurring hazard.

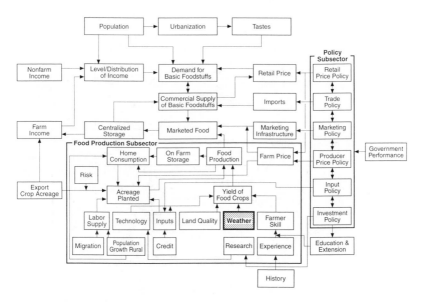

Figure 10.3. Food production diagram for sub-Saharan Africa. It shows the narrowest of views about how food production activities are affected by weather. (From USDA, 1981.)

and other boxes, if one were to replace the word "weather" with the word "drought"? A drought can affect land quality. Drought combined with poor land use practices, especially in semiarid areas, can initiate or accelerate desertification. A drought can affect on-farm storage, labor migration, the use of inputs such as fertilizers, labor supply, farm income, farmers' skill, imports, and so forth. A more accurate portrayal of how weather in the broad sense (that is, both good and bad weather) can affect food production efforts is depicted in the modified USDA chart shown in Figure 10.4.

Any information, such as an El Niño forecast, that can serve to forewarn society of the possibility of a regional drought could be used by governments and farmers alike to modify their climate-sensitive activities as a way to "hedge" their decisions in the event that drought conditions do materialize. In addition, such information can be used by the international humanitarian community to prepare to respond on short notice to food shortages in countries that might require foreign assistance.

1991–92 drought in southern Africa

In early 1992, southern Africa was hit by what many observers have called its worst drought of the century. In the early stages of this drought, it was estimated that a drought-related famine threatened the lives of up to 80

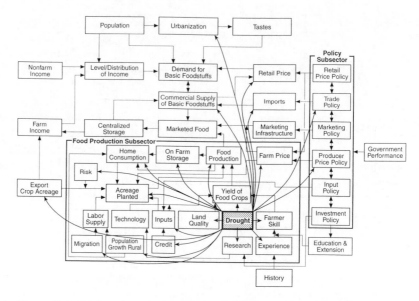

Figure 10.4. Food production chart to account for the way drought can affect food production activities in sub-Saharan Africa. (Modified from Figure 10.3.)

million people in the region. The countries of southern Africa had entered this drought situation on the heels of poor rainfall and crop production during the 1990–91 growing season. For example, commercial farmers in Zimbabwe during this period had diversified their crop portfolios away from maize production, as a partial response to declining prices to producers for their maize. As a result, local pockets of hunger and famine in the region already existed in 1991. Because of this situation, southern Africa's regional early warning unit, as early as March 1991, issued a warning that regional food stocks were dwindling toward such a low level that there would be very little exportable grain within the region.

While these food-related problems were slowly unfolding in 1991, several countries in the region were also involved in structural adjustment programs (SAPs) fostered by the International Monetary Fund (IMF), the World Bank, and other international donors. These SAPs were designed to reduce stagnant economic growth, budget deficits, and inflation. However, they also had direct effects on both regional and national food security. For example, Zimbabwe exported relatively large portions of its reserve maize stocks in the regional marketplace throughout 1991 in response to IMF efforts to reduce government spending for maize storage and to increase its foreign exchange earnings (Stoneman, 1992). Before structural adjustment, Zimbabwe had enough maize reserves to cover at least 6 months of consumption, which would have allowed sufficient time to arrange for additional food needs caused by the 1991–92 drought. Several months later, at the same time that severe drought was setting in, the Zimbabwe government was forced to purchase maize in the international marketplace at prices that were up to three times higher than what it had been sold for a few months earlier.

According to a 1992 report of the Institute of Development Studies in the UK, "the very earliest warning signals at the end of 1991 had been generalized by early 1992, so that by February it was clear that the situation in the region was extremely serious" (IDS, 1992, p. 4). Once information about an impending severe drought-related regional food shortage became irrefutable in the early months of 1992, various national, regional, and international organizations responded over the next several months to improve the regional food supply situation. Their efforts to import food into the region were hampered by the fact that Zimbabwe and South Africa, the region's traditional food surplus areas (and, therefore, regional food exporters), suffered major drought-related crop reductions. Ultimately, widespread famine throughout southern Africa was averted but at a substantial financial cost to national, regional, and international governments and institutions.

Droughts in southern Africa have been associated with El Niño by several scientific researchers. It is unclear, however, to what extent El Niño information was available to, and used by, national and international decisionmakers involved in dealing with the 1991–92 drought. Food security officials in the region reported that they did not receive information about the possibility of an El Niño event until December 1991. Therefore, such information was not used in their regional food-security-related decisionmaking process. However, some El Niño researchers have noted that a reliable forecast of the 1991–92 El Niño event had been available as early as March 1991. Could the response to the food crisis have been better? Could it have been more timely, more efficient, and less costly? Had an earlier forecast of drought been made available to regional policymakers, that information might have been influential enough to generate early positive responses to possible, if not likely, food shortages in southern Africa. Such responses might have saved the countries of southern Africa and the international community hundreds of millions of dollars in costs incurred by unnecessary emergency-level responses. The 1991–92 drought situation in southern Africa raises questions of whether the appropriate decisionmakers received El Niño information and, if not, why not? Given the financial costs of averting famine in southern Africa during the 1991–92 drought, these questions merit further investigation.

The El Niño research community has done a fairly good job of exposing to the public the importance of El Niño research to society. I consider this to be the "wholesaling" of El Niño information. However, it is now time to "retail" that information to potential users. Retailing is a much more difficult and time-consuming task. It involves identifying sector by sector, activity by activity, whether and how El Niño information might be used in specific decisionmaking activities. An improved understanding of El Niño and its related teleconnections can provide a forecast tool that can be used to mitigate the impacts on society of extreme meteorological events such as drought. Such a forecast tool can go a long way to aid decisionmakers in their plans to cope with food shortages, adverse health effects, and other economic impacts associated with recurrent El Niño events.

A forecast is a forecast. It's not a guarantee.
(Yakima (Washington) superintendent, June 1977)

Drought is when the government sends you a report telling you there's no water.
(Yakima (Washington) farmer, July 1979)

11 "In their own words..."

This chapter presents some thoughts of researchers who have been concerned with the El Niño phenomenon during the past several decades. In a way it is a testimonial to the importance of the phenomenon and to the dedication of the scientific community in its attempts to provide society with scientific information that is potentially highly valuable to humankind.

The original intention was to ask only a few researchers, specifically emeritus researchers, to reflect on El Niño or on some aspect of their careers relating to it. My invitations to participate in this chapter were carried out by telephone. Each conversation about El Niño research was highly interesting and prompted me to call "just one more scientist." Before long, the few contributions I originally sought turned into many. And the list would have been even longer, had space permitted it.

The contributors represent different interests and a variety of disciplines from the physical, biological and social sciences. They also present a variety of views, sometimes conflicting, about the prospects of developing a reliable and credible El Niño forecast capability in the near future. They also hold different views on the causes as well as the impacts of El Niño events. Some of them refer to the air–sea interactions in the Pacific as El Niño, while others call it ENSO. Some speak of the predominant importance of sea surface temperatures while others focus on atmospheric circulation. In this regard these paragraphs in their own words can serve as a microcosm of the range of views that exist in the larger scientific community.

None of the contributors had seen the rest of the manuscript at the time they wrote their comments. Thus, I consider their comments to be an integral part of the book, with their views sometimes reinforcing, sometimes supplementing and sometimes challenging points raised elsewhere. No restrictions were placed on their paragraphs, except that they had to relate in some way to El Niño.

César Caviedes

Department of Geography, University of Florida, Gainesville, Florida

Being raised and educated in Valparaiso, on the Pacific coast of South America, for me El Niño was a life experience before it became an empirical scientific concept. In my sophomore year in college, El Niño 1957 left a visual imprint in my memories, as I saw thousands of *guanays* (guano birds) arrive fatigued on the beaches and die of starvation in the Bay of Valparaiso. These images were vivid reflections of a marine-climatological phenomenon about whose technicalities I was to learn several years later reading an enlightening paper by Warren Wooster. For us central Chileans, the El Niño years 1957, 1958, or 1965 meant more rains, more dying *guanays*, less edible fish, regardless of the name of the phenomenon. Actually, nobody, or just a few initiated, knew that these were concomitants of an oceanic anomaly that was in full swing off the shores of northern Peru.

Already a trained geographer, and established in North America, the initial impressions that I had of El Niño began to be shrouded in the vestments that science imposes on true-life experiences: my concept of El Niño had to be expressed in sea surface temperature variations, precipitation anomalies expressed as deviations from certain averages, reductions or increases of the baroclinic fields, Southern Oscillation indices, teleconnection figures, and all the other necessary paraphernalia that now go with the El Niño–Southern Oscillation theme. Number crunchers, modelers, and forecasters took precedence over the witnesses of reality: they were more likely to obtain research funds and have their papers published in "reputable" journals.

Perhaps because from early on El Niño meant to me a human experience, I consistently insisted in bringing to light the human implications of El Niño and felt committed to report on the suffering of coastal Pacific dwellers when El Niño strikes. Engaged in this task, I was thrilled to discover – in 1971 – that, when this phenomenon hits the western coast of South America, in northeastern Brazil, on the other side of immense South America, *secas* (droughts in Portuguese) besiege the rural populations of the *sertão* (the semi-arid zone). More than any other statistical measure, this discovery of another scourge on the human condition was a revelation of teleconnections at their best!

Working on the human-social aspects of El Niño has been a task not exempted from frustration, particularly when I see the collective inability to capture in objective and accurate traits the negative sequels that this climate-ecological crisis inflicts on humble populations, precisely those who receive most brutally the impact of climate variations.

We have reached a stunning level of precision and sophistication in describing and quantifying the numerous parameters of ENSO across the planet. Our capacity to predict these oscillations will probably increase with the help of powerful computing tools and intricate programs in the future. But we are incredibly behind in properly surveying the social effects that ENSO causes to deprived communities. The aggregate amounts of dollars in damages that are quoted whenever El Niño strikes hardly reflects the misery, loss of livelihood, and the traumas experienced by the fishermen and oasis farmers of Peru, the llama shepherds of the Altiplano, the agriculturalists and riverine dwellers of the Paraná River, the herders of sub-Saharan Africa, or the sheep ranchers in remote Australia.

Surely remarkable strides have been made in the scientific understanding of ENSO since the early 1970s, when El Niño started to become a household concept among the general public in North America, but we are still in a pitifully backward stage in what pertains to studying and trying to solve some of the sociological and human problems caused by this phenomenon. We can only hope that the stunning progress achieved in the scientific aspects of El Niño–Southern Oscillation is also paralleled by major advances in the human-social side of this story before the third millennium begins.

Warren Wooster
School of Marine Affairs, University of Washington, Seattle, Washington

El Niño has grown from a geographical curiosity when I first encountered the phenomenon in 1957 to its present recognition as a major climate signal. Certainly the nature of the physical event is now much better understood, and its reliable prediction in the equatorial Pacific is at hand. But foreseeing the tropical event does not foretell its likely biological consequences, even in the tropics. And whether or not an event will have significant extra-tropical effects, and what they will do, for example, to the return of salmon or to the reproduction of bottom-dwelling fish in high latitudes is even less well known. The atmospheric and oceanic circulation in the high latitudes is surely related to El Niño (or, to be fashionable, ENSO) and so the high latitude biological productivity in the ocean must also be linked. Working out the nature of that linkage, to the point of making useful predictions from tropical events is a fascinating scientific enterprise as well as one with considerable utility. It may also be the key to anticipating the response of marine ecosystems to impending climate change.

George Philander

Atmospheric and Ocean Sciences Program, Department of Geography,
Princeton University, Princeton, New Jersey

The El Niño of 1982-83 caught all the experts by surprise. It prompted the rapid development of a capability to predict that phenomenon and, thus, introduced the new era of operational predictions of climate fluctuations. Scientists are now striving to provide routine forecasts, not only of El Niño, but of global climate fluctuations such as exceptionally harsh winters, persistent droughts and prolonged periods of heavy rainfall. This activity has interesting parallels with operational weather prediction.

L. F. Richardson first proposed in the 1920s that the weather be predicted numerically. He envisioned a room with thousands of people making the necessary calculations. Realization of Richardson's dream became possible after the invention of the electronic computer during World War II. The effort started when the mathematician John von Neumann assembled a small group of atmospheric scientists in Princeton, New Jersey, to use the new computer to forecast the weather, and rapidly blossomed into the mature science of weather prediction whose results are now readily available in newspapers and on radio and television, including a 24 hour channel devoted entirely to weather. To keep the public informed of impending storms and the whereabouts of hurricanes requires a global network of measurement platforms that include polar orbiting and geostationary satellites. Several supercomputers assimilate the data and are essential tools for anticipating weather patterns several days hence. The beauty of this activity is the manner in which it integrates the efforts of scientists with diverse interests including experimentalists who design instruments to measure the atmosphere, and theoreticians who develop models of the atmosphere and also of the oceans and land conditions.

At present, climate forcasting is in its infancy; it is where weather prediction was half a century ago. In retrospect the attempts at weather prediction in the 1950s may seem primitive, given the state of computers and of the global measurement network at that time. However, without those efforts, weather prediction would not be as successful as it is today. There will no doubt be criticisms of current efforts to anticipate the next El Niño but without those efforts we are unlikely to make progress. We are at the beginning of a new era that promises rich rewards.

Klaus Wyrtki

Department of Oceanography, University of Hawaii, Honolulu, Hawaii
(emeritus)

Ever since I was a post-doctoral researcher in Kiel, Germany, and observed the wind generation of the seiches [standing waves] in the Baltic, I asked myself the question: do similar seiches occur in the large ocean basins? A study of sea level records in the Pacific finally led me to the conclusion that El Niño was an equatorial Kelvin wave confined to the upper few hundred meters of the ocean. It has been most gratifying to see that my explanation of the dynamics of El Niño as an ocean response forced by the winds has stimulated an enormous amount of subsequent research and the creation of the Tropical Ocean–Global Atmosphere (TOGA) program. This in turn has lead not only to scientific insights into the large-scale ocean-atmosphere interactions, to the development of dynamical models, and to predictive models for El Niño, but most importantly to the establishment of an open-ocean monitoring system. If we ever want to document and understand variations of the climate system, a permanent monitoring system in the ocean is a basic necessity. Process-oriented experiments, favored by many scientists and the funding agencies, are no substitute for such a global ocean monitoring system.

Early on I participated in efforts to predict El Niño, but soon I became convinced that a prediction of the time of occurrence and of the intensity of the next event will have a high degree of uncertainty, because I strongly believe that the ocean–atmosphere system is an inherently non-linear turbulent system which cannot be deterministically predicted a year or more in advance.

David Enfield
Atlantic Oceanographic and Meteorological Laboratory, Physical
Oceanography Division, NOAA, Miami, Florida

As I ponder past progress in understanding ocean–atmosphere interactions and their effects on climate, the most difficult and interesting challenge I see for the near future is how to devise ways of applying that knowledge to benefit global society. How, for example, can we fashion prediction schemes that tailor physically feasible predictions to the pragmatic everyday needs of people and human economic activity? This involves, among other things, getting very different communities to exchange ideas effectively. How do we do that? How can we adapt our present ability to predict sea surface temperatures months in advance to the task of forecasting whether a rainy season will start early or late? How do we tell engineers and insurance underwriters what the future probabilities of strong El Niño events are, years or decades in advance? Climatologists must become statisticians and vice versa; oceanographers must communicate with fisheries experts; vocabularies and jargon must be translated between communities; the general public must be educated; and people of widely differing nationalities and backgrounds must learn to jointly apply themselves to problems that transcend traditional political and disciplinary boundaries. We are only beginning to think about how to do these things. Before they can truly become a reality and impact on society in meaningful ways, the perceptions and consciousness of entire communities must be transformed and an entirely new generation of scientists and decisionmakers must be prepared to carry on with this new paradigm.

Mary Voice
National Climate Centre, Melbourne, Australia

During my career in climate services in Australia, a major hurdle has been jumped by the temperamental thoroughbred named El Niño. El Niño has escaped from the corral of the scientific world and is now running free in the community. Over the past decade, a significant section of the Australian community has heard of El Niño and knows that it is linked to Australian droughts. Many people in climate-sensitive industries have a broad understanding of the Australian consequences of El Niño and also its temperamental behavior. This is a major step forward for a country like Australia, so strongly affected by El Niño.

Like all new theories released into the domain of human communication (the media, word of mouth, etc.), and like that thoroughbred just over a hurdle, we must now keep its head pointed in the right direction. As scientific communicators, we must ensure that every storm and every seasonal fluctuation is *not* attributed to El Niño through false enthusiasm. We must hold onto the reigns and point our thoroughbred El Niño in the direction of cautious utility for the community, rather than the wild speculation of the racecourse betting ring. This need will increase, as more and more "punters" join in the predictions of the future track record of El Niño.

We are at the barricades of a new race, the horses are in the starter's hands and ready to jump. Our thoroughbred El Niño (perhaps with a name change to ENSO) is the clear favorite for delivering scientifically credible and useful seasonal to interannual climate predictions. The task ahead is exciting: to hold our thoroughbred on track, to call the race accurately so that pundits and watchers alike understand the odds clearly, to manage the fluctuating performance that any thoroughbred will sometimes deliver, and to communicate our knowledge to all so that informed decisions win out over wild speculation.

Gerald Meehl

Climate and Global Dynamics Division, National Center for Atmospheric Research, Boulder, Colorado

One of the genuinely gratifying aspects of studying ENSO is that it has such a rich history. There is some (perhaps bitter?) consolation in knowing that the same things confounding you today, as you try to understand ENSO phenomena, are the very things that confounded Walker, Berlage [Dutch meteorologist], Bjerknes, Troup [Australian meteorologist] and others decades ago. A consequence of this long history is that various features are rediscovered from time to time, for example, the seasonal cycle of ENSO. It does not follow the calendar year, and transitions between the warm and cold phases tend to occur in the Northern Hemisphere spring. ENSO has persistence on the annual timescale until a transition occurs, often in northern spring and, not surprisingly, transitions are the most difficult to forecast. This aspect of ENSO has been known for decades, but has recently come into play as something called the "spring predictability barrier." Modeling studies seem to fail most consistently trying to forecast across the northern spring season.

The challenge and excitement of studying ENSO is that we are simultaneously observing, analyzing, modeling, and trying to forecast ENSO behavior from next season to next century. We can take pride in how far we have come in measuring, understanding, and simulating all of these interactions. But with the likes of Sir Gilbert Walker and Jacob Bjerknes looking over our shoulders, we can also see how far we have yet to go.

John M. Wallace

University of Washington, Seattle, Washington

Much of the progress of the past 20 years in understanding El Niño can be attributed to the advances in our ability to observe the tropical atmosphere–ocean system and to recognize the patterns that we see. Many of us share fond memories of three colleagues who made notable contributions to those advances: Verner Suomi, who invented much of the satellite instrumentation that provides today's global perspective on El Niño; Adrian Gill, whose lucid papers on equatorial wave dynamics expanded and unified our vision of the role of the atmosphere in El Niño; and Stanley Hayes who pioneered the concept of an operational ocean observing system in support of El Niño prediction. All three of them were dreamers and incurable optimists, whose bold scientific aspirations have become reality. Were they able to contribute to this volume I am sure that they would be looking forward to even greater achievements as scientists seek to observe, understand, and ultimately predict El Niño, the chaotic pulse of the climate system.

Mike Hall
Office of Global Programs, NOAA, Washington, DC

When scientists began organizing large-scale research efforts to understand El Niño and its relationship to global climate, it became evident that a new school of scientific thought would be needed. The combination of closely interrelated oceanic and atmospheric processes embodied by ENSO in the tropical Pacific required scientific minds not only conversant with, but insightful in, both meteorology and oceanography. Research program managers believed at the time that this fusion of existing disciplines into something new had to be engendered in large part through management practices. Procedures were adopted to encourage research proposals aimed at the coupled problem, for example.

In retrospect, it seems far more appropriate to acknowledge that nature created the new discipline of coupled ocean–atmosphere thinkers by placing before us an intellectually seductive problem whose solution demanded, and thus engendered, the evolution of scientific thought across disciplines. Management practices merely followed nature's lead.

Now as we examine the role of ENSO-forced climate variability in human affairs, we see the need for an even broader evolution of thought. In this case, nature compels us to develop a cadre of skilled economic and social scientists intensely interested in the physical system and how it works, and a cadre of expert physical scientists whose excursions into the behavior of human systems are motivated by more than a casual curiosity. The two should eventually merge into the new discipline we seek. Today's pioneers of this new school of thought will be joined by an increasing number of scientists fascinated by the intellectual challenges which lie at the interface of the two disciplines. Once again a natural evolution of thought will have been driven by an unmistakably important problem.

In building research programs and securing the main resources to design and implement them, we can be confident that nature will provide a clear lead if we are careful to define a workable and relevant problem whose solution requires thinking across disciplines. As always, humans learn when nature teaches.

Timothy P. Barnett
Scripps Institute of Oceanography, La Jolla, California

In the late 1970s, climate research was a poorly thought-of branch of science, and attempts to predict climate were generally looked upon as thoroughly disreputable. So, in 1979–80, when Klaus Hasselmann and I used some fancy statistics to show El Niño events off Peru, we expected to take some heat. No one could shoot us down technically, so they merely dismissed the results. The same methods were used to show that the huge 1982–83 event was predictable a year in advance. At that point, a few science program managers in Washington, DC, began to listen; they saw the prediction ability as justification for future large programs. Indeed, EPOCS and TOGA were based on the promise of predictive skill.

During the mid-1980s, two other groups used dynamical or "hybrid" techniques to advance the results of the earlier statistical studies, which themselves were substantially improved. These very diverse approaches to prediction all said the same thing: moderate to large ENSO events are predictable at lead times of roughly a year or more. These claims, which were again treated with disbelief by many, were backed up by real-time, public forecasts covering the period from 1986 to the present. Perhaps the most interesting of these forecasts was the call for the first major cold event in 15 years to occur in 1988. This forecast was made during a talk in the autumn of 1987, at the height of a major warm event. The listening audience was highly skeptical. But the success of that forecast, and the numerous others made since by several methods, leaves little room to doubt our ability to predict big ENSO events. The real questions that remain are (a) "How far in advance can we predict?" and (b) "How can we say in advance that the forecast will be a good (or poor) one?"

Perhaps the most gratifying part of the forecast business is our recently demonstrated ability to take the long-range forecasts of equatorial Pacific water temperatures and use them to force an atmospheric model, thereby making global climate predictions up to a year in advance, or longer. This approach has shown substantial skill, not only in the tropics, but also over midlatitude regions of North and South America, parts of Asia, and Australia. So, in less than 15 years, we have gone from statistical forecasts of water temperature off Peru (that most folks put no faith in) to a situation where we now routinely make skillful, operational forecasts of global climate with highly sophisticated physical models based solely on our ability to forecast equatorial sea surface temperatures. The latter forecasts are being made today with a variety of statistical and dynamical models.

It has been an exciting ride to have been so closely involved in this rapid advancement, especially since it will have a huge benefit to humankind.

Vernon Kousky
Climate Prediction Center, National Weather Service, NOAA,
Washington, DC

Since the 1982–83 ENSO, the most severe ENSO this century, real-time climate monitoring and climate prediction techniques have evolved to the point where reliable one- to three-season forecasts of conditions in the tropical Pacific are routinely being made. These forecasts are being used extensively in regions where ENSO has a significant relationship with precipitation and/or temperature in order to optimize economic planning. Still, much remains to be done in applying ENSO predictions for economic planning purposes and in scientific research on climate prediction techniques. What are the practical limits of predictability for the climate system? Can we develop prediction techniques that will skillfully predict the observed episode-to-episode variability and yield information as to the magnitude of individual episodes? These are just some of the challenges facing scientists as we approach the twenty-first century.

Stephen E. Zebiak
Lamont-Doherty Earth Observatory, Columbia University, New York

El Niño has been a key player in the earth's climatic variability for a very long time, but has only very recently become appreciated in scientific, economic, and social terms. Scientifically, the study of El Niño has led us to a genuinely new understanding of the nature and strength of the interplay between the earth's oceans and atmosphere, and the implications this has for all aspects of climate. The recent advances in predicting El Niño, though limited, are exciting, as they pave the way for more applied forecasts that can be put to immediate use in agriculture, water resources, and a myriad of other applications. There is much about climate and its impacts that is not yet understood, but from the study of El Niño it is already possible to foresee new uses of climate information in aspects of human activities that may benefit societies worldwide. It has been, and remains, a pleasure to participate in this research.

Ants Leetmaa
Climate Prediction Center, National Weather Service, NOAA, Washington, DC

We are fortunate to work in a time when nature has given us strong ENSO events (to motivate and provide for our support), a few odd events to provide humility, hard-won observations to limit theoretical speculation, physical models that capture essential real (and imaginary) physics, enough computer power (so we don't have to think too much), to explore new regimes, and good friends to work and compete with. The baby that was born, a fledgling ability to forecast El Niño, is something quite remarkable.

This is not the end, but the beginning of a journey to understand the irregular heartbeats of ever-present climate variability.

J. Shukla
Institute of Global Environment and Society, Calverton, Maryland

El Niño is a good example to illustrate that there is indeed predictability in the midst of chaos. My own interest in the modeling and predictability of El Niño was kindled by a fascinating lecture by Mike Wallace (V. Starr Memorial Lecture at MIT in 1980). El Niño provided the strongest validation of the then-emerging hypothesis that the tropical climate is highly predictable. Although we still have a long way to go in modeling and predicting El Niño with fully coupled ocean–atmosphere models (I do not know of any fully coupled ocean–atmosphere model which has yet predicted an El Niño in real time), the current success with simplified models has already added a new chapter in the history of climate research. The conventional wisdom of the mid-twentieth century – that there is no predictability beyond two to three weeks – must be modified to recognize that the slowly varying boundary conditions at the earth's surface, produced by interaction among atmosphere, ocean, and land surface introduce, at times, a highly predictable component in an otherwise chaotic climate system.

George Kiladis
Environmental Research Laboratories, NOAA, Boulder, Colorado

ENSO research has provided more than just clues to the workings of the ocean–atmosphere system over the tropical Pacific. ENSO is a major recurring natural "experiment" that enables scientists to test their ideas about the physical mechanisms responsible for variability of the global system. The experiment consists of disruptions of the atmosphere during warm and cold events in the Pacific Ocean. Most of the solar energy intercepted by the earth does not heat the atmosphere directly, but instead evaporates water from the surface of the ocean, which later condenses in clouds, releasing the "latent" heat that was stored as water vapor. In this regard tropical thunderstorm activity can be viewed as the primary "heat engine" responsible for driving atmospheric motions, even as far away as the polar regions.

During an ENSO event, the thunderstorm activity is altered in intensity and location and the atmospheric circulation responds by adjusting to the new location of the energy source. Thus, the eastward shift of thunderstorns from Australasia into the central and eastern Pacific during El Niño should have some consequences, such as a stronger jet stream and an altered storm track over the North Pacific. Since these expected changes are actually observed to take place, and can also be modelled in computer simulations, this carries much weight in favor of our current "understanding" of how the atmosphere works.

Each new ENSO that is better observed using more advanced technology can provide us with improved data to understand the complicated structure of the climate system. It seems safe to say that ENSO has provided the motivation to improve our understanding of many of the processes in the ocean–atmosphere system that are quite basic and fundamental to the maintenance of the earth's climate itself.

Kevin Trenberth
Climate and Global Dynamics Division, National Center for Atmospheric Research, Boulder, Colorado

As a young scientist in New Zealand in the early 1970s, I began exploring sources of the pronounced year-to-year variations in local climate. Analyses of available limited-area data left research questions partly unanswered; the dominant phenomenon proved to be essentially global in scale – the Southern Oscillation. Obtaining global datasets for both the atmosphere and the ocean in order to analyze and understand the El Niño–Southern Oscillation (ENSO) has since been a constant quest. This quest has brought great satisfaction, because various aspects of the phenomenon have been unveiled.

I consider myself very fortunate to have participated throughout the 1985–94 Tropical Ocean–Global Atmosphere program. The great progress in understanding the coupling between the atmosphere and ocean in the tropical Pacific and linkages to weather around the world has brought benefits to society and human activities in many regions. Yet, because ENSO continues to amaze us with its variety of behavior, we must continue to seek to unravel nature's secrets.

Mike McPhaden
Pacific Marine Environmental Laboratory, NOAA, Seattle, Washington

The past decade has seen breathtaking advances in our understanding of, and ability to predict, El Niño–Southern Oscillation events. Two major developments have provided the impetus for these advances. One has been the development of numerical coupled ocean–atmosphere models for studies of year-to-year climate variability. The other has been the development of an ocean observing system for improved description and detection of climate variations. The two have interacted synergistically, with models providing insights into the environmental variables that need to be measured, and the measurements providing data in real time for incorporation into numerical climate forecast models.

As one measure of progress, consider that the 1982-83 El Niño–Southern Oscillation event, the most intense of the century, was neither predicted, nor even detected until near its peak! A little over ten years later, we routinely monitor the pulse of the tropical Pacific (and other tropical oceans) day-by-day from an extensive network of moored and drifting buoys, volunteer observing ships, islands and satellites, and we are making skillful experimental forecasts up to one year in advance. The benefits to society of this scientific revolution, though potentially enormous, are just beginning to be realized. The challenge now to maximize these benefits will require the collective effort of environmental scientists, economists, social scientists and policymakers from around the world developing tailored analysis and prediction products for specific regional applications.

Peter R. Gent
Climate and Global Dynamics Division, National Center for Atmospheric Research, Boulder, Colorado

The El Niño–Southern Oscillation is the largest year-to-year feature of the Earth's climate. During the last ten years, this oscillation has been intensively observed in the Tropical Ocean–Global Atmosphere program. Great progress has been made in modeling this phenomenon, so that predictions of the future evolution of tropical Pacific sea surface temperatures are now being made routinely. Progress, however, has been slower in the understanding of how this oscillation affects the atmosphere away from the tropical Pacific.

In order to capitalize on this understanding, further progress is required on two fronts. The first is continuing the observation, modeling and prediction of the oscillation and its connections to the global atmosphere. The second is to understand how a reliable El Niño–Southern Oscillation forecast can be utilized effectively. This is more straightforward in countries surrounding the tropical Pacific, but less so in countries, such as the USA and Canada, which are more distant and where the oscillation's effects are more varied.

Harry van Loon
Climate and Global Dynamics Division, National Center for Atmospheric Research, Boulder, Colorado

Two points about the Southern Oscillation which were stressed long ago by scientists studying it have influenced me. They are worth repeating.

(a)   In 1957, H. P. Berlage of the Royal Netherlands Meteorological Institute came to the conclusion that

> anything useful which was achieved in [inter]seasonal forecasting was arrived at by application of the Southern Oscillation, and furthermore... the Southern Oscillation is no pseudo-periodic fluctuation, but a physical process of worldwide extent, by which the general circulation in both hydrosphere and atmosphere is accelerated and decelerated rhythmically with a period varying between 2 and 3 years.
>
> (Berlage, 1957, p. 152)

The fact that the cold and warm extremes of the Southern Oscillation are superposed on a two to three year oscillation has been re-demonstrated in the past few years. And it is obvious that any success in interseasonal forecasting over the Indian and Pacific Oceans and the adjoining continents stems principally from persistence in the Southern Oscillation.

(b)   With regard to using the correlations in the Southern Oscillation system for long range forecasting, Sir Gilbert Walker issued the following warning in 1936:

> predictions can only be issued with restraint if public confidence is to be won. The natural consequence is silence, except when the indications are markedly favorable or unfavorable. In a race with 30 starters a conspicuous good horse may, without undue risk, be backed to come within the foremost 6, and we may feel confident that a thoroughly bad animal will be in the last 6. It may at first sight seem a confession of weakness to issue no forecast when conditions appear roughly normal; but it is better to admit your limitations, and only speak when you can do so with some safety than to issue predictions when they are little more than guesses.
>
> (Walker, 1936, p. 130)

With a little twist this statement holds for public forecasts based on physical models as well. Namely, when you don't know the complete physical structure of the phenomenon – and we don't, being constantly surprised by new observations – you should restrain yourself to improving your model to such an extent that a public forecast has a fairly high probability of success.

Peter Webster
Department of Atmospheric and Oceanic Studies, University of Colorado, Boulder, Colorado

It is clear that the study of the joint interannual variability of the coupled ocean–atmosphere system has demanded the creation of a new breed of scientist, one who is equally adept in understanding processes in the ocean and in the atmosphere. This is an important change in the paradigm of scientific concentrations that have channeled our thinking for many centuries. However, this new amphibious beast has more responsibility than merely understanding common vagaries of two spheres instead of one. The amphibian is at the forefront of a revolution in the very fabric of human endeavor and humankind.

For the first time, humankind may have the ability to plan economically, agriculturally, and societally beyond the realms of the annual cycle. The ability to project from a knowledge of the vagaries of ENSO that the summer to come will not merely be warmer than winter (a projection based on knowledge of the annual cycle), but that it will be very much warmer and perhaps even wetter than average, is a formidable tool. Such knowledge releases humankind from a dependence on planning from one year to the next and allows it the freedom of multiyear planning. Just as the knowledge of the annual cycle allowed the development of the agrarian society some 5000 years ago, humankind stands at the edge of the development of an ability to optimize the full range of societal ambitions. Thus, while past research tethered us to the annual cycle, the ability to forecast variability from one year to the next portends a potential that is unprecedented. With these advances, however, come new responsibilities. Climatologists must now take into consideration the implications for society of long-term forecasts. For the first time, perhaps, weather scientists have the opportunity to work in parallel with social scientists and planners.

Henry F. Diaz
Environmental Research Laboratories, NOAA, Boulder, Colorado

Great strides have been made in the last 15 years in understanding the El Niño phenomenon. This improved understanding has translated into more accurate and reliable predictions of El Niño occurrences, and of the far-flung influences on the climate in many areas of the world engendered by this quasi-periodic warming of the tropical Pacific Ocean. However, one is reminded of the story of Prometheus and the gift of fire: a tool of great promise to humanity, but one with a sharp double edge! It is vitally important that we learn how to use our improved knowledge wisely.

Antonio Busalacchi
Laboratory for Hydrospheric Processes, NASA/Goddard Space Flight Center, Greenbelt, Maryland

Over the past 30 years, our knowledge of El Niño research has progressed from a relatively obscure phenomenon off the coasts of Ecuador and Peru, to a topical item of discussion at cocktail parties, to a level where experimental forecasts of El Niño and the Southern Oscillation are being used for the tangible and practical benefits to society in tropical countries that span the globe. The opportunities for an oceanographer or meteorologist to contribute to the entire progression of a particular problem from basic research, to applied research, and ultimately to improvements in the quality of life for humankind are few and far between. It has been a privilege to be associated with the cadre of dedicated scientists working on the El Niño problem, but with this privilege there is also the responsibility to see this line of research through to the point where experimental forecasts of El Niño are sustained and made routine in a manner that benefits both developing and developed countries around the world.

Mark Cane
Lamont-Doherty Earth Observatory, Columbia University, New York

When I started in oceanography, it seemed to me to be very interesting, but essentially useless. So it is a surprise to be where I am now. Skill in predicting ENSO impacts is not very high, but is already at a useful level. Much interesting work remains to translate what we can say about climate variability into usable forms. Ultimately, it may be that the major importance of ENSO prediction is in establishing the idea that aspects of climate are predictable, and that these predictions can be put to some use.

Richard Barber
Duke University Marine Laboratory, Beaufort, North Carolina

El Niño has taught two lessons that will endure. The first is that large-scale variability such as El Niño is not a disaster, anomaly, or cruel twist of fate; it is how Earth works. To mature and live harmoniously in the Earth system, human culture must adapt to Earth's rhythms and use natural variability to its advantage. El Niño is an integral component of the El Niño–Southern Oscillation cycle that determines the character of a large portion of the Pacific Ocean and the meteorological character of much of the world. The El Niño phase of this cycle is required for the non-El Niño phase to exist. The rich fisheries of the eastern boundary of the Pacific, for example, are the products of the cycle. The lean, warm years are the price the Pacific pays for the fat, cool years. Plants and animals of the eastern Pacific are adapted to this variability. After massive El Niño-driven decreases in plankton, fish, birds, and mammals (the world's largest natural biotic abundance cycle), these populations respond with massive reproductive increases. The harmony of this physical-biological oscillation is the first lesson we must take to heart. A side issue biologists will add is that each passage of the cycle exerts awesome natural selection; each El Niño must increase the fitness of the organisms to exploit the variability cycle. No wonder our Coastal Upwelling Ecosystems Analysis (CUEA) program had as a basic tenet the idea that upwelling ecosystems are the evolutionary cauldron that brews more fit species for less variable parts of the world ocean.

The second lesson is a human one. El Niño–Southern Oscillation variability is the first great coupled atmosphere–ocean–biota puzzle that humankind has solved. This knowledge has value in its own right, but as a symbol of what is possible it has greater significance; it tells humans that they can know how Earth works. Extracting this understanding is not easy. It takes many governments with vision and good will, creative individuals sharing their best ideas to reach a common goal, and ordinary bureaucrats, engineers, scientists, and technicians working with selfless dedication and little recognition. Marshaling all those requirements might sound unrealistic today, but in fact that was what happened in the last two decades when the El Niño puzzle was solved.

Colin Ramage
Department of Meteorology, University of Hawaii, Honolulu, Hawaii (emeritus)

Most of us have no trouble recognizing a moderate or strong El Niño with hindsight. It is such a massive and prolonged event that we are hypnotized into believing that it can be predicted. But we tend to forget that every El Niño comprises a collection of individual synoptic events (e.g., warm or cold fronts), that may be favored by preconditions established in the equatorial Pacific, but which are by no means guaranteed by them.

Apparently, a succession of atmospheric surges from the middle latitudes, of preferably the Northern Hemisphere, generates surface eastward-flowing wind surges and, hence, oceanic Kelvin waves in the eastern equatorial Pacific and so cause El Niño. Early surges usually sweep equatorwards just east of the Philippines or Australia. After the first equatorial westerly wind burst, feedback between the equator and higher latitudes may enhance the *chance* of a succession of surges, but does not guarantee them. Surge frequency probably varies annually, tending to favor El Niño starting early in the year and lasting about a year. Recent observations, however, suggest that (depending on how El Niño is defined) a long enough record would contain starts and stops in any of the months and durations from six months to more than two years. Forecasting El Niño demands that such surges be predicted. I would suggest that success in forecasting El Niño events is slightly more likely than in synoptic (weather) forecasting.

Chester Ropelewski

Climate Prediction Center, National Weather Service,
Washington, DC

Society has been interested in seasonal climate prediction since humans first started to depend on agriculture for survival. In recent decades climatologists have come to understand the physical processes and impacts of one important part of the climate system, the El Niño–Southern Oscillation (ENSO). Conceptual models of the physical components of ENSO have formed the basis of numerical computer models. Progress in climate prediction is considered by many scientists to be analogous to that achieved in weather analysis and forecasting. In weather forecasting, the conceptual framework of "frontal" theory early in the twentieth century was followed by the development of experimental numerical forecasts in the 1950s and, in turn, has evolved into the routine multi-day weather forecasts of today. Based on this analogy, potential users of climate forecasts have an expectation of receiving routine, accurate, and useful computer-model-based multi-season climate forecasts in the not too distant future. Only time and experience will tell whether this optimism is warranted and whether society has finally realized dreams initiated centuries ago.

Stefan Hastenrath

Department of Atmospheric and Oceanic Sciences, University of Wisconsin,
Madison, Wisconsin

From the analysis of observations of rainfall anomalies in key tropical regions, an understanding has been reached of the circulation mechanisms that are operative in regional climate anomalies. On this (empirical-diagnostic) basis, methods of climate prediction have been developed for some regions (for example, for Brazil's Nordeste).

The Southern Oscillation (SO) is also reflected in changes of the large-scale atmospheric circulation. It is, then, not surprising that changes of the Southern Oscillation will have some association with variations in rainfall in many regions. However, only a few of these relationships are strong enough to be useful for prediction.

The Pacific El Niño phenomenon occurs during what is considered the low Southern Oscillation phase, due to forcing by the atmosphere. Subsequently, the warm water anomalies, covering much of the tropical Pacific Ocean, feed back on the global atmosphere. In this context, it is noteworthy that in India summer monsoon rainfall is most strongly associated with the phase of the SO in the following, rather than in the preceding, months.

Experience in tropical climate prediction over the past ten years has revealed a multitude of feasible targets for prediction and a diversity of useful methods. The application of numerical models to climate prediction has concentrated on the Pacific El Niño phenomenon with some success. Widespread publicity, however, may leave the unfortunate impression that this is the only meaningful target and approach. In reality, a broad-based effort is called for, directed to various target regions, combining empirical-diagnostic approaches with numerical modeling methods.

James O'Brien
Center for Ocean–Atmosphere Prediction Studies, Florida State University, Tallahassee, Florida

It is a shame that almost everyone blames bad events such as floods, droughts, fish kills and heat waves on El Niño. El Niño is a good dude! When the waters are warmed along the equator from the dateline to Ecuador by 1 deg.C or more above normal for more than five months, the condition is called El Niño. It occurs every three to seven years. The regional climates over all the Pacific Rim countries are affected. For example, major changes occur in fisheries. In the late 1970s, scholars blamed the loss of Peruvian anchovies on El Niño. We now know that the ocean currents along the western coast of South America, during El Niño, go south. The wonderful shrimp along the Ecuadorian coast can be caught off Peru and the Peruvian anchovies can be caught off northern Chile. These countries now take advantage of the El Niño-related climate shifts in the location of various fish populations.

For the southeastern USA, El Niño is wonderful. It brings gentle, extra rain in autumn and winter, which in the dry season suppresses forest fires from Arizona to Florida. In the Atlantic Ocean, the number of hurricanes are suppressed. The probability of two or more hurricanes reaching the USA in a regular year is almost 50%. When El Niño occurs in the winter before the hurricane season, the probability drops to less than 20% that two or more hurricanes will strike the USA.

The opposite of El Niño is El Viejo (the old man) which means water cooler than average by 1 deg.C or more. When equatorial water is cool, there are more forest fires across the southern USA and more hurricanes occur in the Atlantic. The production of winter crops such as oranges are reduced, unless the rain deficit is somehow replaced. Recent published studies have documented that at least $US100 million per year savings can be realized by farmers in the southeastern USA through crop rotation and the changes in timing of planting using reasonably successful El Niño forecasts. It is now possible to predict the onset of strong El Niño events about a year in advance. Unfortunately, we only predict the sea surface temperature along the equator in the tropical Pacific Ocean. Predictions for agriculture, fisheries, forestry, droughts and floods are projected from previous occurrences. In the future, coupled models of the ocean and atmosphere will predict these recurring climate shifts in the Pacific.

Gary D. Sharp
California State University, Monterey Bay, Seaside, California

Despite decades of anecdotal tales and profound public statements about pending events (that never materialize), or of great oceanographers having stomped out of El Niño workshops muttering "there will be no El Niño ..." at the onset of the 1982–83 event, claims in the quest for a credible ENSO forecast are routinely forfeited as the tally of "surprises" mounts. There are other atmospheric processes, known and unknown, within which the ENSO cycles are intertwined, such as an internal oscillation of the earth's atmosphere, called the Quasi-Biennial Oscillation, or QBO.

Colorado State University atmospheric science Professor William Gray has long pointed out that, as economically devastating as the 1982–83 warm event has proven to be, the ENSO cold phase consequences could make the costs associated with warm events seem trivial. In addition, the benefits of the impacts of the *entire* ENSO cycle (both warm and cold events) on the periodic revitalization of marine and terrestrial ecosystems has certainly been undervalued.

Environmental science is young. There is not yet a sufficient observational basis for statistical utility. Nor is there much real understanding about the range of normal climate variation, except within the paleoclimate community. We must continue to expand our perceptions as we grow older and wiser. Look upstream, look downstream, high and low ... and learn.

William Gray
Department of Atmospheric Sciences, Colorado State University, Fort Collins, Colorado

The strength and frequency of El Niño events have exhibited variability across several decades. El Niño episodes were more prominent during the first two decades of this century and again since the late 1960s than in the intervening decades. In both of these periods the Sahel region of Africa experienced severe drought conditions, major Atlantic hurricane activity was greatly depressed, and the North–South Atlantic Ocean thermohaline circulation (part of the global ocean's conveyor belt circulation; Figure 11.1) appears to have been weaker than normal.

Gilbert Walker described and defined the Southern Oscillation primarily during what in retrospect was an active era of El Niño activity (1900–1920). But ENSO activity lessened in the 1920s to 1960s in an

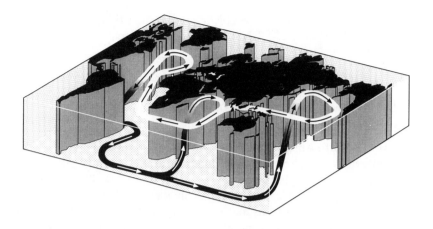

Figure 11.1. The great ocean conveyor belt, depicting global thermohaline circulation. Colder water in the North Atlantic sinks to the deep ocean, to resurface and be rewarmed in the Indian and north Pacific oceans. Surface currents carry the warmer stream back again through the Pacific and South Atlantic. This circuit takes almost 1000 years. (From Broecker, 1987, illustration by Joe LeMonnier.)

apparent response to a stronger Atlantic Ocean conveyor belt. His pivotal research was largely ignored, until the Atlantic Ocean conveyor belt slowed up in the late 1960s and we began seeing more frequent and stronger El Niño events. I predict that, when the Atlantic Ocean conveyor belt speeds up again in coming years, we will once again see a downturn in ENSO strength and frequency and a consequent slackening of interest in ENSO as the primary (or only) seasonal climate determinant. Then we will begin to better appreciate that there are other primary short-term climate "drivers" such as the wind shifts in the stratosphere, extra-tropical Pacific basin features, and so forth, all of which are important climate players.

Richard W. Katz
Environmental and Societal Impacts Group, National Center for Atmospheric Research, Boulder, Colorado

The repercussions of ENSO-related research for other fields are not very well appreciated. For instance, Gilbert Walker's work on the Southern Oscillation and its teleconnections resulted not only in the Walker Circulation being named after him, but also involved an original development in statistics known as the Yule–Walker recursion (a technique employed by Walker to model the quasi-periodic features of the Southern Oscillation). Few are aware that both these labels refer to the same

"Walker." Today analogous problems require contributions from disciplines such as statistics, if ENSO forecasts are to realize their potential value to society.

Carlos Nobre
Brazilian Space Research Institute, São José dos Campos, Brazil

Undoubtedly the ENSO phenomenon strongly affects the regional climates of South America. Statistical or numerical seasonal predictions of droughts in northeast Brazil or of floods in southeastern South America rely heavily on atmospheric and oceanic conditions of the Tropical Pacific. This knowledge has been used by decisionmakers for the benefit of society, as is the case of the use of seasonal forecasts to guide agricultural policies in northeast Brazil with a measurable gain in yields. However, by the same token that no two ENSO episodes are alike, large climate anomalies can occur in the tropics and subtropics for which the causes cannot be directly attributed to ENSO. For example, in 1995 subtropical South America east of the Andes mountains experienced its warmest winter of the last 15 years. In the past, very warm winters had always been associated with a warm tropical Pacific Ocean. Such was not the case in that year, and the research community has been forced to look for the causes of such a large and persistent climate anomaly. This has raised some disturbing questions about the actual short term predictability of the tropical climate.

Pablo Lagos
Peruvian Geophysics Institute, Lima, Peru

During an El Niño event, abnormally warm ocean surface temperatures appear across the equatorial Pacific and along the Peruvian coast, giving rise to anomalous climatic patterns over wide areas of the globe. The floods and droughts that tend to occur during major El Niño episodes are frequently catastrophic.

An improved understanding of the El Niño phenomenon over the past decade has enabled scientists to forecast the onset of El Niño episodes up to a year in advance. Climate, when we can anticipate its behavior, constitutes for the end-users of climate information, a new natural resource that can be used in planning and decisionmaking to increase productivity in the main economic sectors. It is the application of El Niño forecasts that affords society an opportunity to turn scientific information into a useful tool to mitigate its adverse impacts and to undertake efficient socioeconomic planning activities in order to improve the quality of human life.

Neville Nicholls
Bureau of Meteorology Research Centre, Melbourne, Australia

ENSO has short-term effects (such as a drop in crop yields) but also long-term, sometimes permanent effects. Long-term effects arise because of the way that humans react to the short-term climate anomalies caused by an ENSO event. For instance, a drop in fish numbers because of an El Niño, combined with overfishing throughout the duration of an El Niño event, could lead to the destruction of a fish population. In Australia, keeping cattle stocking rates at high levels as we go into an ENSO-related drought, can lead to permanent degradation of the grazing lands. However, not restocking quickly when good rains return at the end of the drought can (and has in the past) result in the spread of "woody weeds"; forests of shrubs and small trees replace good grazing land. Even destroying the habitat of indigenous animals or changing the natural fire regime can lead to long-term effects from subsequent El Niño events. Even with perfect predictions of El Niño and El Niño-related short-term climate fluctuations, it would be difficult to predict and avoid some of these long-term consequences, because of their close dependence on human actions and reactions. The avoidance of such long-term consequences will be a major challenge as we improve our El Niño predictions. And other long-term changes, such as a changing global climate, will make this avoidance even harder to achieve than it might be today.

Wang Shao-wu
Department of Geophysics, Beijing University, Beijing, China

I have been working since 1954 in the field of climatic change and long-range weather forecasting. However, the cool summers in 1969, 1972, and 1976 changed my research focus. These low temperatures in summer brought tremendous damage to crop yields in northeast China. Michael Glantz and I had a very interesting talk about this more than ten years ago in Nairobi. He mentioned to me the possibility of the linkage between the cool summers and ENSO, sparking my interest in this phenomenon.

During the past ten years, four long climate series have been reconstructed: cool summers in East Asia, the number of landed typhoons in China, the drought/flood index in north China, and finally, ENSO. Each of these time series is about 500 years long, enabling me to study the relationship between variations in climate in China and historical aspects of ENSO. Research on ENSO and its impacts on the climate of China has taken up a great part of my scientific life. I never thought this was a bad choice, and I shall continue this research for the rest of my life.

Eugene Rasmusson
Department of Meteorology, University of Maryland, College Park,
Maryland

The story of how diverse lines of oceanographic and meteorological research converged to reveal the elegant system of physical and ecological interactions now known as the El Niño–Southern Oscillation (ENSO) phenomenon is a fascinating tale of scientific progress. However, the trail of discoveries that finally led to our current understanding of this important feature of climate variability was long and difficult, over a century in length, with many twists and turns, and not a few dead ends and diversions. I was among the fortunate climate researchers to be "at the right place at the right time" in his scientific career to participate in the exhilarating, but often humbling, scientific quest for understanding. This quest followed the general recognitions in the late 1960s of the potential importance of this climate phenomenon for seasonal-to-interannual climate prediction. Time and again during this period, we have felt that "now we understand," only to have Mother Nature reveal a new aspect of this phenomenon which sent us "back to the drawing board."

The last chapters in this story are yet to be written. New discoveries and many surprises are still in store for those who pursue the elusive goal of a full understanding of the ENSO cycle and its application to the prediction of climate variability. For this work we need creative and enthusiastic young scientists who are no more shackled to past explanations than were the earlier heroes of this story, Gilbert Walker and Jacob Bjerknes.

12 Usable science

The scientific community is busy producing considerable amounts of information based on its research efforts. From the perspective of the public, much of that information tends to rest in relatively obscure books and journals and does not often get put to use by people outside that relatively small research community. With the end of the Cold War in the late 1980s and the demise of the Soviet Union in the early 1990s, coupled with budgetary pressures to cut spending, governments have been questioning the value to society of government-funded scientific research. For example, members of the United States Congress have complained about whether to continue support for what they considered to have been curiosity-driven research that has no demonstrable social value, as opposed to basic or applied research. Given that national economies are under stress, governments are seeking ways to cut their budgets and scientific research appears to be an area that has attracted the attention of budget-cutters. Some of their concerns have focused on the notion of "usable science."

In recent years, what constitutes "usable science" has been a focus of attention not only within the United States Congress but within the scientific community as well. In 1990 Congress established the United States Global Change Research Program, part of whose charge was to produce "usable" scientific information. Usable science would better inform decisionmakers on how to respond to global environmental change.

It has become more important than ever to bridge the gap between scientific output and societal needs. We, as a society (individuals as well as government funding agencies), must put more effort and resources into mining the scientific information that is being produced in ever-increasing quantities for use in addressing societal needs and wants.

Under the label of "usable science," the Environmental and Societal Impacts Group at the National Center for Atmospheric Research convened several workshops in the mid-1990s to identify potential uses as well as potential users of El Niño information. The first workshop was held in October 1993 in Budapest, Hungary, and focused on the use of El Niño information in food security and famine early warning. Although the workshop placed major emphasis on agricultural production in sub-Saharan Africa, representatives from Latin America, Central and Southeast Asia were also involved.

The agenda of a second usable science workshop, held in November 1994 in Boulder, Colorado, was directed toward identifying the potential uses and misuses of El Niño-related information in North America. This meeting brought together producers of forecasts, climate impacts researchers, industry representatives, and government decisionmakers in an effort to identify the needs of the user community for El Niño information.

One of the key observations of the participants at this workshop was that, in light of the recent advances in scientific knowledge about air–sea interactions in the Pacific Ocean (for example), it was time for the social science research community to become actively engaged in seeking ways to broaden the use of such information by society. The workshop focused not only on the physical processes of El Niño, but also on identifying those groups in North America that can use El Niño-related information to enrich their decisionmaking capabilities, especially with regard to their climate-sensitive activities.

The third Usable Science workshop was held in October 1995 in Ho Chi Minh City, Vietnam. Its purpose was to bring to the same forum El Niño researchers, climate impacts specialists, disaster relief experts, and Southeast Asian educators. The objective was to convince those in charge of regional environmental education and training programs that a careful use of El Niño-related information could greatly benefit societies in that region. Apparent strong connections between El Niño events and the hydrologic cycle in Southeast Asia countries, specifically the Mekong River, were noted. Linkages were also identified between the dreaded dengue fever outbreaks in Vietnam and El Niño events. Managers of several regional education and training courses for junior and mid-level government decisionmakers agreed to include El Niño information in their courses.

In a November 1995 keynote speech to science ministers from about 50 countries, who were attending, in Washington, DC, the International Forum on Forecasting El Niño, United States Agency for International Development Administrator J. Brian Atwood commented on the influence of these particular "usable science" activities. He remarked that

> The first workshop, held in Budapest in October 1993, focused on El Niño and famine early warning. The result was a landmark article in the scientific journal *Nature*, linking El Niño events to changes in maize yields in Zimbabwe. This result was widely reported in the media, including the *New York Times*. Subsequent research at USAID demonstrated the links between El Niño and drought disasters in Southern Africa and Southeast Asia.
>
> As a result of this work, USAID's Office of Foreign Disaster Assist-

ance, Food for Peace, our Africa Bureau, our Famine Early Warning System, and our missions in Southern Africa are all working with our local counterparts to plan for El Niño-induced droughts in the future. Through appropriate planning of agriculture, water management, and health services, the region hopes to prevent disaster and the need for costly relief when drought next occurs. El Niño forecasts are central to this effort.

In the process of preparing for and organizing one of these workshops, we searched in popular magazine articles, corporate records, and newspapers for mention of El Niño. Surprisingly, we uncovered numerous references to El Niño in such publications as *Restaurant Business, Food and Beverage Monthly, Soybean Digest, Retailer's Update, Travel Magazine, Cruises,* and so forth. References to El Niño were also found in financial reports of brokerage firms such as Shearson Lehman Brothers, Inc., Kidder Peabody, Paine Webber, Inc., and Oppenheimer & Co., Inc., focusing on the needs of the fertilizer industry, chemicals, grains, food and commodities, machine tools, tourism, and so forth.

Such El Niño "sightings" in popular and financial publications caused me to reevaluate my feeling that there is a lack of knowledge among potential users of El Niño about how El Niño might affect life in North American society in one way or another. Those who have been concerned about increasing the application of El Niño information to societal decisionmaking may be closer to attaining that objective than they might realize. These "sightings" also underscore the need for a different tactic to increase the use of El Niño information by decisionmakers. They suggest that clusters of existing users can be identified and used as magnet groups to attract other potential users. Whenever El Niño is mentioned, increasing numbers of people tend to listen.

The potential for warning farmers in advance of drought and allowing them to prepare for it – not only in Zimbabwe, but in other regions of the world – brings science, literally, right down to ground level.

C. Rosenzweig, 1994

Until recently, I had believed that our "usable science" efforts were only one of a few such efforts being carried out by a handful of groups. I also thought that it was a concern of fairly recent origin. I was proved completely wrong while visiting an antique fair in Colorado where a dealer was selling old British newspapers and magazines dating back to the early 1800s. There, I came across an issue of *The Penny Magazine*, published 160 years ago in June 1836. The British magazine was published by a scientific group called the Society for the Diffusion of Useful Knowledge! A colleague later reminded me of the motto of the American Philosophical Society of Philadelphia in the 1770s: "the promotion of useful knowledge."

Society has not paid as much attention to, or put resources toward, the use of knowledge as it has the generation of that knowledge – particularly during the Cold War decades, when the production of scientific knowledge flourished. It is time for society to improve its efforts to better connect the production of scientific information to its use by society. El Niño provides a good contemporaneous example of usable scientific information.

One can see that El Niño is not something that happens every once in a while off the coasts of Peru and Ecuador. Many policymakers and much of the public have now moved beyond that image of El Niño. Nor is El Niño just a scientific curiosity. It is now acknowledged to be a powerful naturally recurrent phenomenon that disrupts the usual ways that human activities are carried out, for good and for ill, in many parts of the globe. Only recently have social scientists begun to show an interest in El Niño and its impacts on managed and unmanaged ecosystems and on economies.

El Niño research output falls directly into the category of potentially highly usable scientific information. The notion of usable science suggests that some scientific research output is not usable. However, most scientific research ultimately has potential social value. An important challenge that we face is to realize that potential. The issue of usability can be discussed in terms of at least three questions: Scientific information is usable by whom? By when does it have to be used to be considered usable? Who decides whether it is usable?

By whom?

El Niño information can be used by other scientists who have not been directly involved with El Niño research, as well as by national policymakers and the public. It can be used for decisionmaking purposes or to educate people about El Niño.

Concerned parties such as fishmeal exporters, fruit growers, graziers, farmers, water resources planners, hurricane forecasters, fisheries managers, and hydropower producers have potential needs with regard to information about El Niño for decisionmaking purposes. However, given the existing level of scientific uncertainty surrounding the phenomenon and its various characteristics, not all potential users of El Niño information can as yet use that information in their decisionmaking deliberations. For example, sometimes El Niño events are associated with droughts in southern California (1976–77) and sometimes with flooding (1982–83). Eventually, scientists will be able to distinguish the differences among types of El Niño events that lead to different, even opposing, impacts in the same region. As research progresses, uncertainties will be reduced, confidence in the scientific findings will be enhanced and, as users become more attuned and knowledgeable about the significance of El Niño for their decisions, use of the science will increase.

With regard to the educational aspect, there are numerous lessons that can be learned through the enhanced awareness of El Niño information and its potential usage. El Niño research teaches us about the following:

- the globe as a system of interrelated components,
- local and regional weather and climate and their impacts on ecosystems and on human activities,
- effects of weather and climate impacts as they ripple through society, going well beyond the initial impact (such as second-order effects),
- the value and necessity of multidisciplinary research,
- weather and climate events, even though those events are hazards that cannot be prevented, their impacts can be mitigated,
- the relative capabilities of countries and sectors to cope with extreme meteorological events, and
- the power of natural processes.

El Niño information can also be used to generate humanitarian concerns nationally and internationally. It can be used to teach us about the need for and importance of continuous historical and paleorecords. Finally, as a recurrent natural phenomenon, El Niño events provide societies with an opportunity to understand them and their teleconnections. These are just a few examples of what El Niño research output can do for us, if the gap between science and society is successfully bridged.

By when should scientific research be usable?

This is not an easy question to answer. Some observers recognize the need to support basic research that may have no immediate, readily identifiable application. Others put a higher priority on instant gratification and, therefore, support for only what they consider to be applied science. Still others focus on what has been referred to as curiosity-driven research. El Niño research efforts have included elements of each of these. Each contributes to improving our understanding of the phenomenon.

Figuring out the science of El Niño is not unlike trying to put together the pieces of a jigsaw puzzle. Some pieces are easy to identify – those with the flat edges that form the outside border of the puzzle. Others are identified by the pictures on the puzzle, and they are pieced together forming isolated clusters that by themselves provide a set of autonomous, unconnected pictures. The clusters are not yet attachable to each other or to the outside frame.

With regard to the jigsaw puzzle analogy, there are many pieces whose shapes and colors are not readily identifiable and, therefore, become subjected to trial and error. I would contend that this is where we are at present with El Niño research. To date, El Niño researchers have identified the framework and several clusters of knowledge centered on characteristics of the El Niño phenomenon, such as Kelvin waves, teleconnections, the winds, the warm pool, the Southern Oscillation, and the like. The hard part lies ahead. Over the past few decades we have compiled considerable understanding. However, there is much to be learned before we can say that we have reduced scientific uncertainty enough to make forecasts of El Niño consistently reliable. Recent El Niño events in the 1991–95 period have tended to support the view of those who think that we still do not have all of the El Niño puzzle pieces in hand. More research is needed, particularly on how to make the best use of still-improving El Niño information.

Who decides what is usable?

Information that is considered usable by one person may not be considered usable by another. Even if one were to leave this question to groups of people, they would probably not agree on what scientific information is "usable". As noted earlier, people with varying perceived needs for information will emphasize the importance of different characteristics of that information. Some will seek higher degrees of certainty than others, some will favor timeliness over certainty, and so forth. One point, however, stands out: so long as El Niño research is supported by public funds, researchers have a responsibility to society to ensure that the public and their elected representatives are satisfied with their research activities. This means that effort must be spent *both* on educating the public and policymakers about El Niño, and on educating oneself about the needs and wants of the public.

While one could effectively make the argument that all scientific research – whether perceived to be basic, applied or curiosity driven – ultimately has value to society, governments are under pressure to pay serious attention to balancing their budgets. As a result, scientific programs are being more closely scrutinized and priorities are being set among those programs. I believe that the potential payoff of improved knowledge of warm and cold events in the Pacific is so great to developing and industrialized countries alike (but for different reasons) that El Niño research efforts demand continued moral and financial support from governments around the globe. Those payoffs are likely to be incremental, as opposed to the result of a blockbuster scientific breakthrough. Policymakers should recognize this incrementalism as a highly probable scenario, based on the gaps in our El Niño knowledge highlighted by the unexpected, out-of-phase occurrence of the 1982–83 event and the unexpected behavior of El Niño in the 1991–95 period. For its part, the scientific community must present its research expectations in realistic terms and avoid generating misperceptions about what it can offer in the near term in the way of usable El Niño-related information.

In trying to answer the question of who decides what constitutes usable scientific information, it is important to remember that many *potential* users of El Niño information are *as yet unaware* that knowledge of El Niño may be of value to their activities. Over time, the physical and social science communities, working together, must identify who those users are, what El Niño-related information they might need, and assist them in identifying ways to use that information as an aid to their decisionmaking efforts. As the knowledge about the phenomenon expands and as we become more aware of how to apply that knowledge, the number of potential users who will become actual users will increase.

I once wrote that forecasting El Niño would be science's gift to the twenty-first century. I still believe that. But it will not happen without continued support for research into the science *and* the social science aspects of El Niño. It will be a powerful tool, once it has been developed. It is time to prepare decisionmakers at all levels of society, from the individual to the national governments, for that eventuality.

References

Acosta, José de, 1588: *Historia Natural y Moral de las Indias*, Sevilla. In *Obras del Padre Jose de Acosta*, Biblioteca de Autores Españoles, Madrid, 1954.

Ångström, A., 1935: Teleconnections of climate changes in present time. *Geografiska Analer*, **17**, 242–58.

Arntz, W. E., 1984: El Niño and Peru: Positive aspects. *Oceanus*, **27**, 36–9.

Bacastow, R .B., J. A. Adams, C. D. Keeling, D. J. Moss, and T .P. Whorf, 1980: Atmospheric carbon dioxide, the Southern Oscillation, and the weak 1975 El Niño. *Science*, 210, 66–8.

Bakun, A., 1990: Global climate change and intensification of coastal ocean upwelling. *Science,* **247**, 198–201. Barber, R., 1977: The JOINT-I expedition of the Coastal Upwelling Ecosystems Analysis program. *Deep-Sea Research*, **24**, 1–6.

Barnett, T. P., 1977: An attempt to verify some theories of El Ni̊o. *Journal of Physical Oceanography*, **7**, 633–47.

Berlage, H. P., 1957: Fluctuations of the general atmospheric circulation of more than one year, their nature, and prognostic value. *Royal Netherlands Meteorological Institute Yearbook,* **69**, 151–9.

Bjerknes, J. 1961: El Niño study based on analysis of ocean surface temperatures, 1935 to 1957. *Inter-American Tropical Tuna Commission Bulletin*, **5**, 219–303.

Bjerknes, J. 1966: A possible response of the atmosphere Hadley circulation to equatorial anomalies of ocean temperature. *Tellus*, 8, 820–9.

Blanford, H. G., 1884: On the connexion of the Himalayan snowfall with dry winds and seasons of drought in India. *Proceedings of the Royal Society of London*, **37**, 3–22.

Bouma, M. J., H. E. Sondorp, and H. J. van der Kaay, 1994: Climate change and periodic epidemic malaria. *Lancet,* **343**, 1140.

Broecker, W. S., 1987: The biggest chill. *Natural History Magazine*, **96**, 74–82.

Brown, B. G., and R. W. Katz, 1991: The use of statistical methods in the search for teleconnections: Past, present, and future. In *Teleconnections Linking Worldwide Climate Anomalies*, ed. M. H. Glantz, R. W. Katz, and N. Nicholls, pp. 371–400. Cambridge: Cambridge University Press.

Brown, L. and E. P. Eckholm, 1974: *By Bread Alone*. New York: Praeger Press.

Burton, I., R. W. Kates, and G. F. White, 1993: *The Environment as Hazard*. New York: Oxford University Press.

CAC (Climate Analysis Center), 1995: *Climate Diagnostics Bulletin, December* 1994. Washington, DC: National Weather Service, NOAA.

CalCOFI (California Cooperative Oceanic Fisheries Investigations), 1959: Editor's summary of the symposium. *Reports*, vol. VII, 1 January 1958 to 30 June 1959, pp. 211–18.

Cane, M. A., 1991: Forecasting El Niño with a geophysical model. In *Teleconnections Linking Worldwide Climate Anomalies*, ed. M. H. Glantz, R. W. Katz and N. Nicholls, pp. 345–70. Cambridge: Cambridge University Press.

Cane, M. A., S. E. Zebiak, and S. C. Dolan, 1986: Experimental forecasts of El Niño. *Nature*, **321**, 827–32.

Carrillo, C., 1892: Disertación sobre las Corrientes Océanicas y Estudios de la Corriente Peruana de Humboldt. *Boletines del Sociedad Geográfico Lima*, **11**, p. 84. Microfiche.

Cunha, E. da, 1944: *Rebellion in the Backlands*. Translated from *Os Sertoes* by Euclides da Cunha. Chicago, IL: University of Chicago Press.

Cushing, D. H., 1982: *Climate and Fisheries*, London: Academic Press.

Diaz, H. F. and F. Markgraf (eds.), 1992: *El Niño: Historical and Paleoclimatic Aspects of the Southern Oscillation*. Cambridge: Cambridge University Press.

Eguiguren, D. V., 1895: Estudios sobre la riqueza territorial de la provincia de Piura. *Boletin del Sociedad Geográfico Lima*, **317**, 143–76.

FAO (Food and Agriculture Organization), 1974: Report of the Fourth Session of the Fishery Committee for the Eastern Central Atlantic (CECAF). *FAO Fisheries Report*, **151**, October.

Feldman, G., 1985: Satellites, seabirds, and seals. In *El Niño en las Galápagos: El Evento de* 1982–83, ed. G. Robinson and E. M. del Pino, pp. 125–30. Quito: Fundación Charles Darwin para las Islas Galápagos.

Flohn, H. and H. Fleer, 1975: Climatic teleconnections with the equatorial Pacific and the role of ocean/atmosphere coupling. *Atmosphere*, **13**, 96–109.

FUNCEME (Fundação Cearense de Meteorologia e Recursos), 1992: *Monitor Climatico*, **6**. Fortaleza, Brazil: FUNCEME.

Garcia, R., 1981: *Drought and Man, the* 1972 *Case History*, vol. 1: *Nature Pleads Not Guilty*. New York: Pergamon Press.

Ghil, M. and S. Childress, 1987: *Topics in Geophysical Fluid Dynamics: Atmospheric Dynamics, Dynamo Theory, and Climate Dynamics*. Berlin: Springer-Verlag.

Gill, A. E. and E. M. Rasmusson, 1983: The 1982–83 climate anomaly in the equatorial Pacific. *Nature*, **306**, 229–34.

Glantz, M. H., 1981: The societal value of an El Niño forecast. In *Resource Management and Environmental Uncertainty: Lessons from Coastal Upwelling Fisheries*, ed. M. H. Glantz and J. D. Thompson, pp. 449–76. New York: John Wiley & Sons.

Glantz, M. H., 1982: Consequences and responsibilities in drought forecasting: The case of Yakima, 1977. *Water Resources Research*, **18**(1), 3–13.

Glantz, M. H., 1984: Floods, fires, and famine: Is El Niño to blame? *Oceanus*, **27**(2), 14–19.

Glantz, M. H., 1991: *ENSO and Climate Change*. Report of Workshop held 4–7

November 1991, Bangkok, Thailand. Boulder, CO: National Center for Atmospheric Research.

Glynn, P. W. (ed.), 1990: *Global Ecological Consequences of the 1982–83 El Niño Southern Oscillation.* New York: Elsevier.

Gray, W. M., 1993: *Forecast of Atlantic Seasonal Hurricane Activity for 1993.* Fort Collins, CO: Department of Atmospheric Sciences, Colorado State University.

Greenpeace International, 1994: *The Climate Time Bomb: Signs of Climate Change from the Greenpeace Database.* Amsterdam: Stichting Greenpeace Council.

Halpert, M. S., G. D. Bell, V. E. Kousky, and C.F. Ropelewski, 1994: *Fifth Annual Climate Assessment 1993.* Camp Springs, MD: Climate Analysis Center, National Weather Service.

Hammer, G., 1995: ENSO impacts in Australia. *The ENSO Signal,* 3, p. 5. Silver Spring, MD: NOAA/OGP.

Hansen, J .E., 1990: Physical aspects of the El Niño event of 1982–83. In *Global Ecological Consequences of the 1982–83 El Niño Southern Oscillation,* ed. P. W. Glynn, pp. 1–19. New York: Elsevier.

Harrison, D.E. and M.A. Cane, 1984: Changes in the Pacific during the 1982–83 event. *Oceanus,* 27(2), 21–8.

Houghton, J. T., G. L. Jenkins, and J. J. Ephraums (eds.), 1990: *Climate Change: The IPCC Scientific Assessment.* Cambridge: Cambridge University Press.

IDOE (International Decade of Ocean Exploration), no date: *Prologue.* Washington, DC: National Science Foundation.

IDS (Institute of Development Studies), 1992: *Drought and Famine in Southern Africa.* Proceedings of a conference held on 20 July 1992 at IDS, Sussex, UK.

Jackson, M., 1993: *Galapagos, a Natural History.* Calgary, Alberta: University of Calgary Press.

Jacobs, G. A., H. E. Hurlburt, H. C. Kindle, E. J. Metzger, J. L. Mitchell, W. J. Teague, and A. J. Wallcraft, 1994: Decade-scale trans-Pacific propagation and warming effects of an El Niño anomaly. *Nature,* 370, 360–3.

Jennings, F. D., 1981: The coastal upwelling ecosystems analysis program: Epilogue. In *Coastal Upwelling,* ed. F. A. Richards, pp. 13–15. Washington, DC: American Geophysical Union.

Jordán, R., 1991: Impact of ENSO events on the southeastern Pacific region with special reference to the interaction of fishing and climate variability. In *Teleconnections Linking Worldwide Climate Anomalies,* ed. M. H. Glantz, R. W. Katz, and N. Nicholls, pp. 401–30. Cambridge: Cambridge University Press.

Kerr, R. A., 1982: U.S. weather and the equatorial connection. *Science,* 216, 609.

Kerr, R. A., 1994: Official forecasts pushed out to a year ahead. *Science,* 266, 1940–1.

Kiladis, G. N. and H. van Loon, 1988: The Southern Oscillation. Part VII: Meteorological anomalies over the Indian and Pacific sectors associated with extremes of the Oscillation. *Monthly Weather Review,* 116, 120–36.

Koblinsky, C. J., P. Gaspar, and G. Lagerloef (eds.), 1992: *The Future of Spaceborne*

Altimetry: Oceans and Climate Change. Washington, DC: Joint Oceanographic Institutions, Inc.

Kogelschatz, J., L. Solorzano, R. Barber, and P. Mendoza, 1985: Oceanic conditions in the Galapagos Islands during the 1982–83 El Niño. In *El Niño en las Galápagos: El Evento de 1982–83,* ed. G. Robinson and E. M. del Pino, pp. 91–124. Quito: Fundación Charles Darwin para las Islas Galápagos.

Lagos, P. and J. Buizer, 1992: El Niño and Peru: A nation's response to interannual climate variability. In *Natural and Technological Disasters: Causes, Effects and Preventive Measures,* ed. S. K. Mujumdar, G. S. Forbes, E. W. Miller, and R. F. Schmalz, pp. 223–8. Philadelphia: Pennsylvania Academy of Sciences.

Lau, K.-M. and A. J. Busalacchi, 1993: El Niño–Southern Oscillation: A view from space. In *Atlas of Satellite Observations Related to Global Change,* ed. R. J. Gurney, J. L. Foster, and C. L. Parkinson, pp. 281–96. Cambridge: Cambridge University Press.

McKay, G., and T. Allsopp, 1976: Global interdependence of the climate of 1972. In *Proceedings of the Mexican Geophysical Union Symposium on Living with Climate Change,* Mexico City, May, pp. 79–86.

Mosaic, 1974: The sea turns over. *Mosaic,* **5**(1), 25–31.

Mosaic, 1975: All that unplowed sea. A Mosaic Special Issue: Food. *Mosaic,* **6**(3), 22–7.

Murphy, R. C., 1923: The oceanography of the Peruvian littoral with reference to the abundance and distribution of marine life. *Geographical Review,* **13**(1), 64–85.

Murphy, R. C., 1954: The guano and the anchoveta fishery. In *Resource Management and Environmental Uncertainty: Lessons from Coastal Upwelling Fisheries,* ed. M. H. Glantz and J. D. Thompson, pp. 81–106. New York: John Wiley & Sons, 1981.

NMSA (National Meteorological Services Agency of Ethiopia), 1987: *The Impact of El Niño on Ethiopian Weather.* Report. Addis Ababa: NMSA.

Nicholls, N., 1986: A method for predicting Murray Valley encephalitis in southeast Australia using the Southern Oscillation. *Australian Journal of Experimental Biological and Medical Science,* **64**, 587–94.

Nicholls, N., 1987: The El Niño/Southern Oscillation phenomenon. In *Climate Crisis,* ed. M. H. Glantz, R. W. Katz and M. E. Krenz, pp. 2–10. New York: UN Publications.

Nicholls, N., 1991: The El Niño–Southern Oscillation: Recent Australian research. Melbourne: Bureau of Meteorology Research Centre. (Mimeo.)

Nicholls, N. (ed.), 1993a: *Climate Change and the El Niño-Southern Oscillation.* BMRC Research Report No. 36. Report of Workshop held 31 May–4 June 1993. Melbourne: Bureau of Meteorology Research Centre.

Nicholls, N., 1993b: ENSO, drought and flooding rain in South-East Asia. In *South-East Asia's Environmental Future: The Search for Sustainability,* ed. H. Brookfield and Y. Byron, pp. 154–75. Tokyo, Japan: United Nations University Press and Oxford University Press.

NOAA/OGP (National Oceanic and Atmospheric Administration/ Office of

Global Programs), 1992: *International Research Institute for Climate Prediction: A Proposal.* Silver Spring, MD: IRICP Task Group.

NOAA/OGP (National Oceanic and Atmospheric Administration/ Office of Global Programs), 1994: *A Proposal to Launch a Seasonal-to-Interannual Climate Prediction Program.* Silver Spring, MD: NOAA/OGP.

NRC (National Research Council), 1990: *TOGA: A Review of Progress and Future Opportunities.* Washington, DC: National Academy Press.

Palca, J., 1986: Could this be an El Niño? *Nature,* **324,** 504.

Parry, M. L., T. R. Carter, and N. T. Konjin (eds.), 1988: *The Impact of Climatic Variations on Agriculture,* vol. 2: *Assessments in Semi-Arid Regions.* Dordrecht: Riedel Publishers.

Partridge, I. J., 1991: *Will It Rain?* Brisbane: Department of Primary Industries.

Paulik, G. J., 1981: Anchovies, birds, and fishermen in the Peru current. In *Resource Management and Environmental Uncertainty: Lessons from Coastal Upwelling Fisheries,* ed. M. H. Glantz and J. D. Thompson, pp. 35–79. New York: John Wiley & Sons.

Petterssen, S., 1969: *Introduction to Meteorology,* 3rd edn. New York: McGraw Hill.

Pezet, F. A., 1895: The counter-current "El-Niño," on the coast of northern Peru. *Boletines del Sociedad Geográfico, Lima,* **11,** 603–6.

Philander, G., 1995: Letters to the editor. *Bulletin of the American Meteorological Society,* **76,** 80.

Quinn, W., V. T. Neal, and S. E. A. Mayolo, 1987: El Niño occurrences over the past four and a half centuries. *Journal of Geophysical Research,* **92,** C13, 14449–61.

Ramage, C., 1986: El Niño. *Scientific American,* **254**(6), 76–83.

Rasmusson, E. M., 1984: Meteorological aspects of El Niño–Southern Oscillation. In *Proceedings of the 15th Conference on Hurricanes and Tropical Meteorology,* pp. 17–20. Boston, MA: American Meteorological Society.

Rasmusson, E. M. and P. A. Arkin, 1985: Interannual climate variability associated with the El Niño/Southern Oscillation. In *Coupled Ocean-Atmosphere Models,* ed. J. C. J. Nihoul, Elsevier Oceanographic Series, vol. 40, pp. 697–725. New York: Elsevier.

Rasmusson, E. M. and T. H. Carpenter, 1982: Variations in tropical sea surface temperature and surface wind fields associated with the Southern Oscillation/El Niño. *Monthly Weather Review,* **110,** 354–84.

Rasmusson, E. M. and J. M. Wallace, 1983: Meteorological aspects of the El Niño/Southern Oscillation. *Science,* **222,** 1195–202.

Richards, F. A. (ed.), 1981: *Coastal Upwelling.* Washington, DC: American Geophysical Union.

Ropelewski, C. F., 1992: Predicting El Niño events. *Nature,* **356,** 476–7.

Rosenzweig, C., 1994: Maize suffers a sea-change. *Nature,* **370,** 175.

Sears, A. F., 1895: The coastal desert of Peru. *Bulletin of American Geographical Society,* **28,** 256–71.

Stoneman, C., 1992: The World Bank demands its pound of Zimbabwe's flesh. *Review of African Political Economy,* **53,** 94–7.

Sullivan, W., 1961: *Assault on the Unknown: The International Geophysical Year*. New York, McGraw Hill.

Thompson, J. D., 1977: Ocean deserts and ocean oases. In *Desertification: Environmental Degradation in and around Arid Lands*, ed. M. H. Glantz, pp. 103–39. Boulder, CO: Westview Press.

TOGA (Tropical Ocean–Global Atmosphere) Project Office, 1987: *TOGA: A Project of the World Climate Research Programme*. 2nd edn, 1 February, ITPO-1. Boulder, CO: International TOGA Project Office.

Tomczak, M. and J. S. Godfrey, 1994: *Regional Oceanography: An Introduction*. Oxford: Pergamon Press.

Trenberth, K. and T. J. Hoar, 1996: The 1990–1995 El Niño–Southern Oscillation event: Longest on record. *Geophysical Research Letters, 23*, 57–60.

Trenberth, K. E., H. van Loon, M. Cane, M. J. Wallace, J. Young, O. Brown, G. Rasmusson, P. Webster, and T. Barnett, 1986: Letters to the editor. *Scientific American*, **255**(5), 6.

Tribbia, J., 1995: What the Southern Oscillation is: An atmospheric perspective. In *Usable Science II: The Potential Use and Misuse of El Niño Information in North America*, ed. M. H. Glantz, pp. 18–19. Proceedings of a Workshop held 31 October–3 November 1994 in Boulder, Colorado. Boulder, CO: National Center for Atmospheric Research.

USDA (US Department of Agriculture), 1981: *Food Problems and Prospects in Sub-Saharan Africa*. Washington, DC: US Government Printing Office.

Walker, G. T., 1936: Seasonal weather and its prediction. *Smithsonian Institute Annual Report 1935*, pp. 117–38.

WCRP (World Climate Research Programme), 1995: *CLIVAR: A Study of Climate Variability and Predictability*. WCRP-89, WMO/TD No. 690. Geneva: WMO.

WMO (World Meteorological Organization), 1995: *CLIVAR*. Brochure. Geneva: WMO.

Wooster, W. S., 1959: El Niño. *California Cooperative Oceanic Fisheries Investigations, Reports*, vol. VII, 1 January 1958 to 30 June 1959.

Wyrtki, K., E. Stroup, W. Patzert, R. Williams, and W. Quinn, 1976: Predicting and observing El Niño. *Science, 191*, 343–6.

Solution to crossword (Figure 1.2).

Index

99

)

(   ]